Entomology, Ecology and Agriculture

Studies in the History of Science, Technology and Medicine
edited by John Krige, CRHST, Paris, France.

Studies in the History of Science, Technology and Medicine aims to stimulate research in the field, concentrating on the twentieth century. It seeks to contribute to our understanding of science, technology and medicine as they are embedded in society, exploring the links between the subjects on the one hand and the cultural, economic, political and institutional contexts of their genesis and development on the other. Within this framework, and while not favouring any particular methodological approach, the series welcomes studies which examine relations between science, technology, medicine and society in new ways e.g. the social construction of technologies, large technical systems.

Other titles in the series

Volume 1 *Technological Change: Methods and Themes in the History of Technology*
edited by Robert Fox

Volume 2 *Technology Transfer out of Germany After 1945*
edited by Matthias Judt & Burghard Ciesla

Volume 3 *Entomology, Ecology and Agriculture: The Making of Scientific Careers in North America, 1885–1985*
Paolo Palladino

Other Volumes in Preparation

The Historiography of Science and Technology: Whose History? Whose Science?
Thomas Söderquist

Making Isotopes Matter: F.W. Aston and the Culture of Physics
Jeff Hughes

The Development of Molecular Biology
Harmke Kamminga & Soraya de Chadarevian

This book is part of a series. The publisher will accept continuation orders which may be cancelled at any time and which provide for automatic billing and shipping of each title in the series upon publication. Please write for details.

Entomology, Ecology and Agriculture
The Making of Scientific Careers in North America, 1885–1985

Paolo Palladino
University of Lancaster, UK

harwood academic publishers
Australia • Canada • China • France • Germany • India • Japan • Luxembourg
Malaysia • The Netherlands • Russia • Singapore • Switzerland • Thailand
United Kingdom

Published in The Netherlands by Harwood Academic Publishers GmbH.

Emmaplein 5
1075 AW Amsterdam
The Netherlands

British Library Cataloguing in Publication Data

Palladino, Paolo
Entomology, ecology and agriculture: the making of scientific careers in North America, 1885–1985. — (Studies in the history of science, technology and medicine; v. 3)
1. Scientists — North America — History — 20th century
2. Science — Vocational guidance — North America — History — 20th century
I. Title
502.3'7

ISBN 3-7186-5907-7

Contents

INTRODUCTION

THE SETTING

The year was 1962, the world was in the grip of the Cold War. A very popular writer had just penned the following words into the introduction of what was soon to become a best-seller:

> there was once a town in the heart of America where all life seemed to live in harmony with its surroundings. The town lay in the midst of a checkerboard of prosperous farms, with fields of grain and hillsides of orchards where, in spring, white clouds of bloom drifted above the green fields. In autumn, oak and maple and birch set up a blaze of color that flamed and flickered across a backdrop of pines ... Then a strange blight crept over the area and everything began to change ... There was a strange stillness. The birds, for example — where had they gone? ... It was spring without voices ... The roadsides, once so attractive, were now lined with browned and withered vegetation as though swept by fire. These, too, were silent, deserted by all living things ... No witchcraft ... silenced the rebirth of new life in this stricken world. The people had done it themselves.[1]

This desolation was not the result of the much feared, final nuclear confrontation with the "red menace". It was instead the work of DDT and a whole new generation of organochlorine and organophosphorous insecticides, first introduced during the second world war to fight the "insect menace". The author was Rachel Carson, and the book was *Silent Spring*. Carson's prose was perhaps melodramatic, and it decidedly incensed those who viewed these insecticides, with some reason, as crucially important for both providing Western households with cheap, abundant food, and feeding the increasingly hungry and restive peoples of the lesser developed countries.[2] Nevertheless, the deathly landscape evoked by these introductory lines bore some resemblance to the state of the Cañete valley of Peru during the mid-1950s. Here, the intensive use of DDT to protect the local cotton plantations against the Cotton Boll Weevil initiated a cycle of ever increasing use of this and other insecticides to overcome both the resulting resistant strains of the Weevil, and the outbreaks of secondary noxious species, whose natural enemies had now been completely destroyed by the insecticidal regime. The guiding philosophy that "the only good bug is a dead bug" eventually led to a situation so unmanageable that cotton cultivation had to be halted.[3]

Prior to the publication of *Silent Spring*, however, most North Americans knew nothing about such disasters. If asked about entomology, they probably would have first recalled a childhood trip to a museum of natural history, and then ventured a guess that entomologists were somewhat eccentric collectors of beetles and butterflies, hiding in museums with their bottled beetles whenever they were not running about a field in plus-fours, wildly waving their butterfly nets. Carson told her readers otherwise. Entomologists actually worked to free farmers, fruit and vegetable growers, as well as the lumber industry, from any insect species that seemed to threaten their economic welfare. In other words, entomology was more realistically described by the commonly adopted adjective "economic" than by any appeal to more natural historical concerns. Carson then proceeded to describe how animals, often far removed from the fields, orchards, and forests, where DDT and the other new insecticides had been used for their intended economic end. Such as the majestic bald eagle died or became so incapable of breeding that they were doomed to extinction. Given that most Americans lived under the threat of another distant, invisible, and still more powerful killer, radioactive fallout from atomic testing in the American West and on atolls in the Pacific Ocean, it is perhaps not surprising that Carson's melodramatic lines might have reminded the readers of *Silent Spring* of the post-nuclear scenery evoked a few years earlier by Nevil Shute in his own best-seller, *On the Beach*, and thus of their own impending extinction.[4] Later, these same readers' vocal opposition to aerial spraying campaigns against the Fire Ant, Gypsy Moth, and Boll Weevil was accentuated by resonance with other aerial campaigns in far away Vietnam.[5]

For many citizens of the United States, therefore, the destruction of the bald eagle by DDT stood as a symbol for much more than just the price to be paid for a "thoughtless" commitment to the "control of nature", which Carson had condemned by writing that,

> "control of nature" is a phrase conceived in arrogance, born of the Neanderthal age of biology and philosophy, when it was supposed that nature exists for the convenience of man. The concepts and practices of applied entomology for the most part date from that Stone Age of science. It is our alarming misfortune that so primitive a science has armed itself with the most modern and terrible weapons, and that in turning them against the insect it has also turned them against the earth.[6]

The death of the bald eagle also stood for the readers of *Silent Spring* as a symbol for a technological world increasingly and menacingly outside their control. In fact, Carson suggested that the damage done by DDT, sprayed everywhere without their permission, amounted to a violation of every American citizen's constitutional rights.[7]

During the decade following the book's publication, all these anxieties motivated the organization of, and then lent great impetus to, a politically powerful environmentalist movement, which made Carson into something of a new St. Francis of Assisi. Following its patron saint, this movement sought the salvation of humanity by empowering what the influential ecologist and veteran spokesman against all forms of environmental destruction since the days of *Deserts on the March*, Paul Sears, hailed in 1964 as the "subversive" science of ecology.[8] This book is, at the most superficial level, about the relationship between economic entomology and this scientific savior, ecology.

THE ACTORS

Since the publication of *Silent Spring* thirty years ago, and it now seems a world away, a number of writers have sought to identify the "people" responsible for the terrible situation so vividly described by Rachel Carson by focusing on very different phases in the history of entomology. The historian James Whorton engaged in this project by considering, in his book *Before Silent Spring*, the ambiguous political and institutional responses to the sanitary problems caused by overuses and misuses of arsenical insecticides before the advent of DDT.[9] Frank Graham, on the other hand, has written far more passionately about political developments since the publication of *Silent Spring*.[10] As the editor of the official publication of the Audubon Society, an organization that was at the forefront of a campaign during the 1960s to ban DDT, he was keenly aware of the strong opposition of the agricultural industry to the political agenda of the environmentalist movement that had first coalesced around this campaign. Thomas Dunlap's *DDT: Scientists, Citizens and Public Policy* is a relatively more dispassionate historian's account of this same campaign, which focuses especially on developments in American courts of law and the halls of the United States Congress.[11] A similarly focused account is articulated by the political scientist Chrisopher Bosso in *Pesticides and Politics*.[12] John Perkins, in *Insects, Experts, and the Insecticide Crisis*, seeks instead to identify the "people" responsible for what he calls the "insecticide crisis" by considering the evolution of some important methods of crop protection that did not involve any massive use of insecticides like DDT.[13]

These books, together, conjure the image of an unshakable alliance of chemical manufacturers, agricultural lobbying groups, and politicians from rural America, based on a shared determination to oppose, for the sake of financial gain, any control whatsoever over the use of insecticides. This conspiratorial image is most forcefully and colorfully rendered by the leading ecologist, environmental activist, and author of *The Population Bomb*, Paul Erlich, in his preface to entomologist Robert Van den Bosch's *The Pesticide Conspiracy*. Erlich writes that,

> Professor Van den Bosch, a distinguished scientist and "insider" with long experience in the business of controlling pest populations, lays out the story of stupidity, venality

> and corruption as only an insider can. It's all here: the suppression of research on alternative systems, the sale of the honorable traditions of the Entomological Society of America for a mess of booze, the pressure put on scientists in state universities to suppress results unfavorable to the "pesticide mafia", the disgrace of the Department of Agriculture, the rape of the EPA [Environmental Protection Agency] — the whole tragic story.
>
> *The Pesticide Conspiracy* is a book written by a man who is frankly angry, and you will be angry when you've finished it. But anger is not enough. Concerted political action is required if the desperately needed transition to integrated pest management is to be achieved. With such management, insecticides are used when needed in an ecologically sound mix of techniques that minimize damage to the crop, the environment, the farm worker, the consumer, and the farmer's pocketbook. Until the Agricultural Research Service of the Department of Agriculture can be upgraded and the USDA [United States Department of Agriculture] as a whole brought into the battle against pests (rather than in favor of insecticides), there is no hope of reform. And the quickest way to get to the USDA is by weeding out its overlords in the House Agriculture Committee — many of whom should have retired decades ago for the benefit of the nation. As long as such men — ignorant of ecology but having enormous power over agriculture — remain in office, the agro-ecosystem of the United States will continue to run downhill toward ultimate disaster.[14]

In sum, the history of the insecticide crisis is, according to these writers, a tale of corruption: the American agricultural establishment had sold itself out to the chemical industry. The biologists and political radicals Richard Levins and Richard Lewontin build upon this conclusion to argue more generally, in *The Dialectical Biologist*, that the development and use of the ever more powerful insecticides promoted by the agrochemical industry is nothing but a chapter in the historical transformation of American agriculture into a technologically advanced, capitalist enterprise.[15]

Significantly, with the exception of Perkins, none of these writers asks why American entomologists created insecticides such as DDT in the first place. How was it possible for members of a scientific community that at one time played a central role in the development of American ecology — the late nineteenth century naturalist Stephen Forbes is often portrayed in disciplinary histories as a founding father of both entomology and ecology — to create and very successfully promote the use of such an ecologically destructive weapon? These authors' silence suggests that they largely agree with Carson's view that the insecticide crisis was simply the result of the chemical industry having "put poisonous and biologically potent chemicals indiscriminately into the hands of persons largely or wholly ignorant of their potentialities for harm".[16] While there is considerable truth to this view, the claim I wish to advance in this book is that entomologists, even the most ecologically sophisticated ones, were not quite the victims of a terrible, corporate conspiracy, but active participants in the evolution of the insecticide crisis. Not only did they exploit existing institutional formations to advance their goals, they also helped to shape them in ways that both created and foreclosed professional opportunities for their successors to pursue possible alternative relationships between entomology and ecology. Ironically, this can be shown by considering the evolution of the technology for crop protection mentioned by Erlich, in his preface to Van den Bosch's book, as the solution to all the problems caused by insecticides like DDT. This technology was also firstly, and most famously, deployed during the late 1950s to restore order in the devastated Cañete valley. This was "integrated control", and its later development "integrated pest management".

The leading proponents of this integrated approach were some of Van den Bosch's colleagues in state agricultural experiment stations in California and research institutes of the Canadian federal department of agriculture, all thoroughly opposed to the philosophy that "the only good bug is a dead bug". They proposed to control insect "pests", and other biological enemies of the farmer, by taking advantage of the natural mechanisms usually regulating the abundance of all biological populations. More specifically, they claimed that the abundance of these pests could be reduced firstly by establishing effective "biological control", that is to say, by introducing or increasing populations of species that preyed upon or parasitized them. After this, they aimed to time insecticidal treatments according to the life cycles of all the insect species involved, so as to minimize any destruction of the predators and parasites (alternatively, they proposed to use insecticides specifically designed to destroy only the targeted pest).

Widespread adoption of this integrated strategy, according to these Californian and Canadian entomologists, would eventually reduce very substantially the need for any insecticides. During the 1960s they succeeded in evoking a great deal of interest in the approach. Many of them had helped Carson on the more technical points about entomology and ecology which she had made in *Silent Spring*. As a fishery biologist conversant with the ideas of most contemporary ecologists, she readily accepted their belief that the development of their more ecologically sophisticated approach would greatly help to resolve the crisis she had publicized. Integrated control was then presented in *Silent Spring* as one of the most promising alternatives to DDT. In 1965, the President's Science Advisory Committee lent its weight to the claim by recommending, in its report on *Restoring the Quality of our Environment*, the wider adoption of the approach because it would halve the amount of insecticides then used.[17]

Twenty years later, however, the entomologist, leading environmental activist, and former member of the committee that prepared the presidential report, David Pimentel, wrote that the use of insecticides had actually grown quite steadily since the 1960s, and that insects still destroyed about the same proportion of the total agricultural output.[18] This negative assessment was in part the outcome of discussions between a number of entomologists and ecologists (including Pimentel) attending a conference held in 1985 in San Antonio, Texas, on the relationship between ecological theory and integrated pest management. The entomologist Marcos Kogan complained that throughout the 1960s integrated pest management had been extolled as a triumph of "applied ecology", but it was never clear to him what this meant. Its development, he wrote, was largely detached from contemporaneous developments in ecological thinking. He summarized this problem by asserting that,

> ecological theory does not stand for integrated pest management as particle physics stands for the development of practical uses of nuclear energy. Ecology is not an exact science, and integrated pest management often advances ahead of the establishment and, at times, despite the contradictions, of current ecological theories. Yet integrated pest management has been hailed as one of the most powerful intellectual constructs of recent applied biology. The power of this construct stems precisely from the recognition that ecology is integrated pest management's ideological root. Consequently, the perception that integrated pest management lacks a solid theoretical background is a matter of grave concern.[19]

Disregarding for the time being the presupposition of an unproblematic relationship between particle physics and nuclear engineering, this passage suggests that many other entomologists and ecologists before Pimentel and Kogan had believed that there was an indisputable, if difficult, link between ecology and integrated pest management. This difficulty must have been particularly aggravating because it was generally supposed that, in the past, entomologists such as William Robin Thompson and Alexander John Nicholson had contributed very greatly to the advancement of both theoretical ecology and related practices such as biological control. The problem all entomologists and ecologists had to solve was, then, how to reconcile the highly abstract notions about the structure and regulation of ecological communities elaborated by contemporary ecologists with the complexity encountered every day by entomologists manipulating the ecology of insects on agricultural crops. In fact, Pimentel suggested on the same occasion when he reviewed the limited achievements of integrated approaches that more "basic research on the ecology of pests" was needed to transform the problems described in *Silent Spring* into memories from a distant past.[20]

I wish to argue, however, that Pimentel, Kogan, and many other entomologists and ecologists beside them, were simply redefining the problems of economic entomology, and merging them into those of theoretical ecology. In so doing, they were in fact helping to increase the gap between ecology and work of crop protection, a gap which they viewed as chiefly responsible for the insecticide crisis. Uncovering the complex dynamics that engendered this situation will help to clarify what I mean by saying that these entomologists and their predecessors actively contributed to the making of the scenery described by Carson. Hopefully, it may also cause us to reflect more critically on the link between particle physics and nuclear engineering which Kogan cited as the paradigmatic example of the direct dependence of advanced technical practice on the most prestigious contemporary science.

AGRICULTURE AND THE HISTORY OF SCIENCE

The history of science and technology are today inextricably connected. This is chiefly because, during this century, governmental institutions have become the most important sources of funding for science, and have become so largely because such funding is believed to advance the technological changes needed to maintain military and economic pre-eminence.[21] Historians of the physical and engineering sciences are well aware of this situation, and by studying its manifold ramifications have advanced quite considerably our understanding of the relationship between science and modern society.[22] For example, various contributors to Peter Galison and Bruce Hevly's *Big Science* have shown how the post war, large scale institutional organization of research in physics has shaped our knowledge about the world.[23] However, few historians have noted that this was not the only way to organize such research.[24] Fewer still have considered what might be the implications of heterogeneity in the organization of research for the careers and ideas of the often invisible scientists involved in shoring up, rather than leading, these great intellectual transformations. This is, nonetheless, much more than has been done by most historians of the life sciences, who have not ventured much in this domain of "big science" and technological development. Why, for example, is there no essay on massive enterprises such as the International Biological Program or the War on Cancer in Galison and Hevly's volume?[25]

Philip Pauly has recently suggested that the historiography of subjects that have attracted much attention among historians of the life sciences, such as genetics and eugenics, might have to be greatly revised if these subjects were placed more squarely within the context of much broader and very important historical problems, such as the politics of public health and the organization of medical care — another area of major expansion in post war science.[26] This is not simply a matter of illustrating the importance of this medical context for the institutional development of particular biological sciences, which has already been done very insightfully by Robert Kohler for the history of biochemistry, but also of considering the consequences of this development for the subsequent history of both these sciences and the institutions harboring them.[27] In agreement with Pauly's critical perspective on the historiography of the life sciences, I am proposing to look at another very important locus of social and economic activity which has played a major role in the development of these sciences: agriculture.[28]

Agriculture has received very little attention among historians of science and technology. The following few data readily suggest, however, that this makes little sense if we want to a have a fuller understanding of the modern history of these two subjects and their relationship. In his more recent study of American private foundations and their support for the natural sciences, Kohler suggests that during the 1920s the Rockefeller Foundation, which was at that time one of the most important patrons anywhere in the world of the life sciences and their applications to advance human welfare, decided to avoid any involvement in the world of American agriculture because "the vast investment in agricultural science by the federal and state governments made it an unappealing area for private foundations".[29] Kohler's assessment is amply confirmed by the work of the economic historians David Mowery and Nathan Rosenberg on the investments of the American federal government in science and technology.[30] There can be little doubt that until the 1950s, when the National Institutes of Health began their explosive institutional expansion, the United States Department of Agriculture was the major and unrivaled patron of biological research. It was also a reservoir of skilful translators of such biological research into technological and industrial innovations: medical scientists may be widely viewed as having won the second world war by inventing penicillin, but, as Robert Bud has pointed out, the far from straightforward task of showing, to these scientists and the pharmaceutical industry, how to bring large quantities of penicillin to the battle-field was the work of scientists in the Northern Regional Laboratory of the United States Department of Agriculture.[31]

Governmental organizations such as the Department of Agriculture were not, however, the only sponsors of biological research relating to agriculture. David Hounshell and John Kenly Smith have written in their comprehensive history of research and development in the DuPont Corporation that "after the war, *Barron's* identified agricultural chemicals as one of the most profitable areas in the already highly touted chemical industry".[32] This was soon translated at DuPont into a rapid expansion of facilities for biological research. Significantly, the development of DDT (and the synthetic herbicide 2,4 D) were very important in bolstering this confidence about the future of agricultural chemicals. Many of these chemicals are now being supplanted by the products of the biotechnology created during the 1980s by an increasingly closer alliance of research departments in universities and corporate giants such as DuPont, Monsanto and Union Carbide, an alliance which in land-grant colleges and universities goes back to the 1920s and '30s. The reason for this historical alliance is that selling insecticides is a seriously big business: American farmers, in 1991, spent $1 330 million on insecticides alone, and $4 100 million more on herbicides and fungicides.[33]

This link between agriculture and the life sciences was far from from a peculiarly American development. The debt of Canadian biologists to agriculture and its associated facilities for agricultural research is as large, if not larger than that of their colleagues south of the border.[34] Governmental structures, however, are not the same every where, even those of the United States and its much overlooked northern neighbor, Canada. As Douglas Verney has pointed out,

> the visitor to North America who sees expressways, roadsigns, automobiles, and airports may be forgiven for concluding that Canada is merely a photocopy of the United States's original design. This may be true in large measure of Canadian civilization, but it is less true of its culture.[35]

I would extend this case beyond the realm of "culture" to science and technology. Political authority and its institutional expressions are more decentralized in the United States than in Canada. Given what Joseph Ben David has written about the relationship between science and different forms of bureaucratic organization, it does not seem implausible that this difference may have resulted in quite different relationships between science and government in these two countries.[36] What significance should we attach to these possibly diverse relationships? What was their impact on American and Canadian agriculture? And what about impacts on the sciences themselves?

Thanks to the work of the historians Barbara Kimmelman and Deborah Fitzgerald, and the sociologist Jack Kloppenburg, it is clear today that botanists and zoologists employed by the United States Department of Agriculture, the land-grant colleges and universities, and the state agricultural experiment stations, played an important role in transforming the political economy of agricultural production.[37] Although it has not sparked as much controversy among historians and other analysts, the history of Canadian agriculture seems to have been similarly shaped by agricultural scientists.[38] Much less is known, however, about the significance of this alliance of science and agriculture for the history of the sciences to which these American and Canadian scientists dedicated their lives. For example, in his encyclopedic, intellectualist history of ecology, the ecologist Robert McIntosh rarely mentions that many of his predecessors articulated their ideas in agricultural (or otherwise economic) institutional contexts.[39] Sharon Kingsland, in her more critical account of this same history, does take notice of this context, but does not think that it merits more than a passing comment.[40] Yet, there is some evidence that these links had very significant repercussions on the conceptual development of ecology (and genetics).[41] Ronald Tobey, for example, has suggested that the competition in late nineteenth century America between amateur botanists and their rivals in the agricultural experiment stations for authority over the emerging field of agricultural botany motivated the development of a more mathematically rigorous study of plant community ecology (the same may be said for the extraordinary popularity of some novel theories of heredity in those same experiment stations).[42]

Thus, beside its importance for our understanding the dynamics that shaped the emergence of modern life sciences, the history of entomology and ecology in an agricultural context can shed light on questions of some considerable interest to historians of science and technology generally. How do scientists and their institutional sponsors interact? What are the implications of their interactions for both the intellectual development of the sciences involved, and for the establishment of the technical practices associated with these sciences? These are the more general questions which this book (on the relationship between entomology and ecology) seeks to address.

AMERICAN AND CANADIAN ENTOMOLOGY

This book started as a fairly circumscribed study of a debate during the late 1950s between entomologists in the state agricultural experiment stations in

California and the research institutes of the Canadian federal department of agriculture about the importance of ecological theory in the development of integrated control. I wanted to know more about the context that engendered the debate, and what happened to it later. Thus, in the first two chapters, I will consider the institutional development of entomology in the United States and Canada from the late nineteenth century onwards — I have chosen the year 1885 for the title and rough opening date because it was around then that governmental institutions for agricultural research were formally organized in these two countries. I will pay particular attention to the significance of their different organizational arrangements for the changing importance attributed by the entomologists in these institutions to the study of insect natural history (and later, ecology). It is only in this manner that we can begin to have some feeling for the context in which the Californian and Canadian entomologists sought, from the 1950s, to advance and promote the development of integrated control.

This will be to a considerable extent a comparative exercise, in which I will often use the experience of American entomologists as a standard. It is not, however, a study in "national styles of science" such as Jonathan Harwood's study of genetics in Germany and the United States, but of science in different national contexts.[43] More importantly, the implicit subordination of the Canadian experience should be viewed critically. Following Robin Winks, suggestions with regard to histories of Canada, any deviations of the Canadian experience from the American standard should not be taken simply as matters of fact, but should also be taken as possible issues for future criticism of the fairly conventional history of American entomology presented in the first chapter of this book.[44] Furthermore, many of the biographical details and institutional dynamics were clearly shared by entomologists on both sides of the forty-ninth parallel. While I hope, therefore, that these first two chapters will also be read as complementary, illustrating different aspects of entomology in North America generally, this is bound to lead to some ambiguity about what exactly I am contrasting. Sometimes this will strengthen my claims about the institutional shaping of scientists' work and ideas, for what better evidence can one ask than having two entomologists, one American and the other Canadian, with shared training in an American university, but occupying different institutions, again, one American and the other Canadian, producing different ideas and technologies? On the other hand, this ambiguity will sometimes remain unresolved, leaving open questions about the effect of these American and Canadian entomologists' different ideas on the establishment of different institutions. I wish the ambiguities to stand as they are, as invitations to the reader to engage in a future dialogue about the relationship between scientists, scientific ideas, institutions, and technological practices.

In the first chapter, we shall see that in the principal institutions associated with the development of American entomology, the land grant colleges and associated state agricultural experiment stations, this science was radically transformed. Because chemical insecticides suited so well the needs of an agricultural community which the entomologists in these institutions were bound to serve, entomology, originally understood as the study of the natural history of insect species, noxious and otherwise, became synonymous with the development of better insecticides to control insect pests. By the 1950s, American entomologists had come to adopt the chemist, rather than the zoologist or biologist, as their professional role model.

Most Canadian entomologists, like their counterparts in the United States, also found their principal employer in agricultural research centers. These centers, however, were controlled by the federal department of agriculture; as in other Canadian federal institutions, the personnel staffing them was both geographically and institutionally removed from any contact with its nominal constituency. Therefore, the direction of work in these agricultural research centers was not determined by local agricultural communities, as would have been the case in the American state agricultural experiment stations and land grant colleges, but by the scientists themselves, or by administrators very close to them. In such a context, I will argue in the second chapter, entomologists reaped greater professional rewards from advancing, for example, knowledge about the ecology of insect species, than from the development of better methods to deliver chemical insecticides against specific species on specific crops. Canadian entomology followed, therefore, the contemporary path of most other organismic sciences, disappearing into biology.

BIOLOGICAL CONTROL, INTEGRATED CONTROL, AND ECOLOGY

From the late nineteenth century to the 1940s and '50s, biological control, that is to say, the purposeful introduction of insect species preying upon, or parasitizing insect pests, sometimes offered a quite attractive alternative to the use of chemical insecticides. In fact, programs in biological control prospered in the Division of Biological Control of the Citrus Experiment Stations at Riverside, California, and in the Dominion Parasite Laboratory at Belleville, Ontario. In the third chapter, I will discuss at some length the development of these two institutions up to the first proposals for an integrated approach during the late 1950s, and will pay especially close attention to the different importance assigned in Riverside and Belleville to research in insect ecology (while discussing the history of the Division of Biological Control, I will rely very heavily on Richard Sawyer's detailed doctoral thesis on the subject).[45] This, along with the two preceding chapters, will provide the background necessary to understand why the entomologists in Riverside (and later, Berkeley) viewed the relationship between ecological research and the development of integrated control very differently from those in Belleville.

The definitions of integrated control they articulated were distinguishable mainly by the different emphasis each placed on the role of biological controls.

This may have been due to their proponents' opposing ideas about the regulation of animal populations in nature, one of the most controversial subjects of theoretical population ecology ever since the entomologists Robin Thompson and John Nicholson defined in a concise and precise manner its central problem: the relationship between predation, parasitism, and other environmental regulatory factors. The entomologists in Riverside and Berkeley viewed predators and parasites as playing a fundamental and dominant role in determining the abundance and distribution of the prey. This view, they argued, was amply confirmed by the successful record of biological control, a record which further strengthened their virtually unlimited confidence in the effectiveness of this approach to crop protection. The Canadian entomologists believed instead that the very same data on biological control cited by those in Riverside and Berkeley confirmed their own, quite different view of the ecology of insect populations. They believed that the prey's abundance was not determined by just its predators and parasites, but by its total biological and physical environment. They doubted, therefore, the effectiveness of biological control as a widely applicable, fully autonomous pest control strategy, and did not believe that chemical control could ever be replaced by biological control. The two approaches should indeed be integrated with one another, as was sometimes proposed by the entomologists in Riverside and Berkeley, but insecticides would almost always be the mainstay of any such integrated programs.

At the same time, the Canadian entomologists attempted to close a debate engendered by these differences during the 1960s by impugning the competence of their opponents to articulate any meaningful ecological argument. They did so by charging that the views of the entomologists in Riverside and Berkeley were shaped by these entomologists' critical need for immediate solutions to the practical problems confronting them, rather than by any serious ecological considerations. As such, the Californian's views (and practical solutions) were bound to be highly unreliable. I will take up this line of argument, and show that it was not at all clear where the boundary between the latter's concern to develop effective programs and more academic interest in a very important ecological question lay. I will also suggest that the Canadian entomologists' own institutional environment led them to adopt an academic approach to the problem of relating ecology and the work of integrated control, which left little room for an equally reasonable, but more empirical approach to ecological theory.

ENVIRONMENTALISM, ECOLOGY AND INTEGRATED PEST MANAGEMENT

It will be important to bear in mind during the course of this history that, while they bitterly disagreed on all manners of ecological questions, the entomologists in Riverside, Berkeley, and Belleville were at one in regarding the future improvement of integrated control programs as resting on greater institutional support for research in insect ecology. More significantly, by the mid-1960s, it was evident to all of them that the transformation of integrated control into a viable alternative to the environmentally disruptive uses of chemical insecticides so decried in *Silent Spring* would not succeed within the usual institutional networks supporting agricultural research. New institutional formations were needed. However, having learned what was the proper purview of entomology in an institution where little emphasis was placed on the operational development of agricultural science and technology, most of the Canadian entomologists had little taste for this major political endeavor. The practical development of integrated control was left therefore to the entomologists in Riverside and Berkeley.

In the fifth chapter, I will examine how American congressional and presidential politics from the early 1960s to the late '70s offered great opportunities for this last group of entomologists to transform their idea of integrated control into a multi-million dollar national program for the development of integrated pest management into major new ways to protect agricultural crops. In particular, I will do so by challenging John Perkins' contrast between the development of this approach and another alternative to contemporary practices of crop protection which had been proposed by entomologists in the Agricultural Research Service of the United States Department of Agriculture. He has focused on what he identifies as the opposing ideological foundations of the two programs to explain the institutional success of the former. I argue instead that these different histories are better understood as the result of a competition for funding between two groups of scientists, shaped by a complex interplay of environmentalist ideology, the politics of neo-federalism, and the economic crisis of the 1970s. It is only within this broader context that we can fully appreciate how scientists skillfully create institutional environments conducive to the establishment of their scientific and political visions.

INTEGRATED PEST MANAGEMENT AND ECOLOGICAL THEORY

Notwithstanding the increased support for integrated pest management during the 1970s, by the early '80s it was still unclear what role ecology played in the approach. In the last, sixth chapter, I explore once again the relationship between the two. This time, however, I look at the function of mathematical modeling in ecology and integrated pest management, showing how all the ambiguities about defining ecology in the 1960s and '70s, which have already been highlighted in considerable detail by the historian Joel Hagen, were reflected in the contemporaneous development of integrated pest management.[46]

During the 1960s, some of the Canadian entomologists whose work I considered in the fourth chapter turned to "systems theory", a new approach to study complex, cybernetic systems such as those supposedly encountered by ecologists, to articulate a new, holistic view of the behavior of biological populations, and thus establish themselves as quite prominent ecologists, rather than invisible agricultural scientists. Their counterparts in Riverside and Berkeley also relied on this approach because it seemed to provide them with useful tools to examine and resolve the very complex practical problems confronting them in, for example, the cotton fields of the Central Valley. These radically different considerations were easily accommodated within the ambiguous conceptual corpus of systems theory. In fact the two groups cemented an alliance between them in the context of the International Biological Program, a program often equated with the wedding of systems theory and ecology consummated in the birth of systems ecology.

The 1960s and '70s were also the heyday of environmentalism, and the American government sought to respond to the concerns of an increasingly vocal and politically powerful environmentalist movement — American government was responsive government, except perhaps where Vietnam was concerned. Between 1968 and 1974 it provided the recently established International Biological Program with funds previously undreamed by most ecologists, principally because the latter promised to demonstrate under its aegis the capacity of systems ecology to address seemingly intractable problems such the environmental damage caused by DDT. Many of the Canadian and Californian entomologists considered in the earlier chapters were involved in establishing and advancing this International Biological Program, and in promoting the establishment of the national integrated pest management project under its umbrella. They found thereby a place among richly funded academic ecologists. Hitching their wagon to systems ecology, however, entailed costs as well, for theoretical population ecologists in leading American universities soon challenged the intellectual importance of systems theory for the development of ecological theory and its technical applications in entomological practices such as integrated pest management.

Unlike systems ecologists, these theoretical population ecologists argued that the analysis of the complex behaviors of natural populations should be reduced to simple interactions between the individuals of various species constituting any ecological community. Their criticism was aimed specifically at two of the Canadian entomologists, Kenneth Watt and Crawford ("Buzz") Holling. They drew attention of both the various criteria these two used to construct ecological models, and to the relationship they defended between models and data collected in the field to support them. They then proceeded to argue that the use of systems theory to develop ecological theory was illegitimate, because it was the tool of engineers simply interested in precise prediction for the purpose of resolving specific practical problems rather than discovering the general patterns of nature. Indeed, Watt and Holling had been closely associated with agricultural institutions, and, like other systems ecologists, had argued that their new approach to ecology would result in improved strategies for managing natural resources such as forests, fisheries, grasslands, or agricultural crops. Theoretical population ecologists believed, however, that it was far too early to expect any ecologists to possess the sort understanding of ecological systems necessary for such practical tasks. They therefore downgraded systems ecologists from leaders of a supposedly new and powerful science to mere engineers working with an inadequate conceptual framework, and cast in doubt the soundness of long-term public support for systems ecology.

This claim drew virtually no response from academic systems ecologists; the only response came instead from Berkeley, from entomologists developing the models needed for integrated pest management programs. Interestingly, by this criticism of theoretical population ecology, and defense of systems theory in ecology, these entomologists effectively abandoned their earlier views on regulation of animal populations. In the 1950s, they had argued against the theoretical perspective of their counterparts in Canada, but then seemed to ally themselves with them in the mid-1960s. Both the Canadian and Californian entomologists had used systems theory to advance their different lines of work, but whereas the former turned to it for a new way of understanding how ecological communities were organized, the latter had appeared to rely on systems theory simply because it satisfied their very practical needs. Now, however, some of the Californian entomologists viewed this same theory as a very powerful tool for both understanding and manipulating the organization of ecological communities — a practically and politically expedient commitment to systems theory had become part of their intellectual heritage.

There may have been an institutional explanation for this controversy over the place of systems theory in ecology. Unlike systems ecologists, most theoretical population ecologists decided that they would not commit themselves to deliver any practical solutions to the ecological crisis motivating government support for the International Biological Program. Consequently, they failed to benefit from the expanded opportunities for the development of ecological research programs in the late 1960s. This impelled them to articulate what they perceived to be some fundamental inadequacies of systems ecology. The situation for the researchers in Riverside and Berkeley was quite different. Starting in the mid-1970s, land grant colleges responded to widespread criticism of agricultural research as academically uninspiring by accelerating the pace of a reorientation, which had already begun in leading institutions such as the University of California during the 1950s and '60, toward greater emphasis on academic rather than practical achievement. These changes may have been responsible for the efforts by researchers at Berkeley to assert their identity as ecologists. Their stance on the relationship between entomology and ecology appeared to change as their audience did; initially they were seeking to stabilize their institutional niche in the world of agriculture, but then had to seek a new one within ecology. The cost of this may have been a renewed alienation of entomology and ecology, and the continued growth of farmers' dependence on chemical insecticides.

It seems, then, that the anxieties over the relationship between ecological theory and the practice of integrated pest management expressed in 1985, a century after the opening of this history, by David Pimentel, Marcos Kogan, and many other entomologists and ecologists beside them, were due, at least in part, to the fact that, throughout the 1960s and '70s, there was no single notion of what constituted the methodological foundations of ecology. In other words, just as in Andrew Pickering's world of particle physics, there was no agreement on the criteria that should guide the construction of ecological theory, nor any agreement over the significance of empirical data and prediction for validating such theory.[47] Thus it would have been very difficult to establish any uncontroversial notion of the proper relationship between ecology and the work of crop protection. Much more significantly, however, it may be that entomologists' and ecologists' conflicting views on both these crucial foundational issues and the relationship between the two disciplines may be explained in institutional terms. That is to say, these views may have been causally linked to the entomologists' and ecologists' different institutional locations, and the debates between them may have been motivated by important changes in the organization of their research programs. Had their institutional situations been different, these debates may never have taken place. Yet, it is also the case that in both the debates considered here, the participants repeatedly focused on a single issue: how is the abundance and distribution of animal populations determined? The repertoire of possible answers, as we will see, was limited by a shared history of ecological theory and by deeply rooted cultural presuppositions, such as the existence of a "balance of nature".

This history of entomology and ecology may therefore be of some interest to contemporary entomologists and ecologists. It draws attention to some institutional roots of the radically divergent ideas held by researchers in the two disciplines, and to areas of conceptual agreement that allow many entomologists and ecologists to consider themselves members of a single and distinct group of scientists. Divisions between, and within, scientific disciplines are not simply matters of intellectual disagreement that can simply be settled by calling scientific congresses and by argument in learned journals. Relationships between ideas are relationships between the living people who hold them. This is why I have used, and will continue to use, the everyday names of the people involved. I will also use the common names of the insects with which they had to deal, as this is how they referred to them, often acknowledging thereby the essentially economic importance of these insects — the Flour Beetle is something more than just than population ecologists' *Tribolium confusum*. The alliances between the people working with these insects, and hence, between their ideas, are determined by the networks of institutions in which they participate, ranging from ever evolving and changing scientific disciplines to the governmental and industrial organizations that sometimes support these disciplines' growth. Thus, to understand the reasons for the difficult relationship between integrated control, integrated pest management and ecology, we must look at the institutional networks within which the entomologists in California and Canada (as well as those entomologists and ecologists attending the meeting in San Antonio) moved.

Such a study may be of still greater interest to historians and sociologists of science by providing some insight into the relationship between intellectual and social factors in shaping the evolution of scientific disciplines. It seems that to understand the relationship between integrated control, integrated pest management, and theories of ecology, we must attend closely to what entomologists and ecologists have had to say about it, how their explanations relate to both the intellectual histories of their disciplines and the histories of the institutional contexts in which their discussions took place, and how these social dimensions are embedded in the intellectual heritage of the next generation of entomologists and ecologists. In other words, science is indeed, as Bruno Latour puts it, "politics by another name", but, just like politics, it is not a game where everything goes.[48]

NOTES

1 R. Carson, *Silent Spring* (Boston: Houghton Mifflin Company, 1987), pp. 1–3.

2 See N.E. Borlaug, *Mankind and Civilization at Another Crossroads* (Rome: Food and Agriculture Organization, 1971).

3 See T. Boza Barducci, "Ecological consequences of pesticides used for the control of cotton insects in the Cañete valley, Peru" in M.T. Farvar & J.P. Milton (eds), *The Careless Technology* (New York: Natural History Press, 1972), pp. 423–38.

4 N.N. Shute, *On the Beach* (New York: William Morrow, 1957). On the links between *Silent Spring* and fears about nuclear war, see R.H. Lutts, "Chemical fallout: Rachel Carson's *Silent Spring*, radioactive fallout, and the environmental movement", *Environmental Review* 9 (1985): 211–25.

5 See F. Graham, *Since Silent Spring* (Boston: Houghton Mifflin, 1970), pp. 206–10.

6 Carson, *Silent Spring*, p. 297; see also T.R. Dunlap, *DDT. Scientists, Citizens, and Public Policy* (Princeton: Princeton University Press, 1981), p. 96.

7 Carson, *Silent Spring*, pp. 12–3.

8 See P.B. Sears, "Ecology: a subversive subject", *BioScience* 14 (1964): 11–3. On Sears, see D. Worster, *Nature's Economy: The Roots of Ecology* (Cambridge: Cambridge University Press, 1985), p. 233.

9 J. Whorton, *Before Silent Spring: Pesticides and Public Health in pre-DDT America* (Princeton: Princeton University Press, 1975).

10 Graham, *Since Silent Spring*.

11 Dunlap, *DDT. Scientists, Citizens, and Public Policy.*

12 C.J. Bosso, *Pesticides & Politics. The Life Cycle of a Public Issue* (Pittsburgh: University of Pittsburgh Press, 1987).

13 J.H. Perkins, *Insects, Experts, and the Insecticide Crisis. The Quest for New Pest Management Strategies* (New York: Plenum Press, 1982).

14 P. Erlich, "Preface", in R. Van den Bosch, *The Pesticide Conspiracy* (New York: Doubleday, 1978), pp. vi–vii, quote from p. vii. On Paul Erlich and

environmentalism, see D. Fleming, "Roots of the new conservation movement", *Perspectives in American History* 6 (1972): 7–91, on pp. 52–7.

15 R. Levins & R.C. Lewontin, *The Dialectical Biologist* (Cambridge: Harvard University Press, 1985), pp. 207–24 and 238–41.

16 Carson, *Silent Spring*, p. 12.

17 President's Science Advisory Committee, *Restoring the Quality of our Environment* (Washington: United States Government Printing Office, 1965), pp. 227–91.

18 D. Pimentel, "Is *Silent Spring* behind us?" in G.J. De Marco *et al.* (eds), *Silent Spring Revisited* (Washington: American Chemical Society, 1987), pp. 173–87.

19 M. Kogan, "Preface", in Kogan (ed.), *Ecological Theory and Integrated Pest Management Practice* (New York: Wiley, 1986), pp. ix–xi, on p. x.

20 Pimentel, "Is *Silent Spring* behind us?", p. 183

21 D.C. Mowery & N. Rosenberg, *Technology and the Pursuit of Economic Growth* (Cambridge: Cambridge University Press, 1989).

22 See, for example, P. Forman, "Behind quantum electronics: national security as basis for physical research in the United States, 1940–1960", *Historical Studies in the Physical and Biological Sciences* 18 (1987): 149–229.

23 P. Galison & B. Hevly, *Big Science: The Growth of Large-scale Research* (Stanford: Stanford University Press, 1992).

24 See A. Baracca, "Big science vs. little science in postwar physics" in M. De Maria *et al.* (eds), *The Restructuring of the Physical Sciences in Europe and the United States, 1945–1960* (Singapore: World Scientific Publishing, 1988), pp. 150–60.

25 On other problems of the historiography of "big science", see J.H. Capshew & K.A. Rader, 'Big science: Price to the present', *Osiris* 7 (1992): 3–25.

26 P. Pauly, "Essay review: the eugenics industry — growth or restructuring?", *Journal of the History of Biology* 26 (1993): 131–45.

27 R.E. Kohler, *From Medical Chemistry to Biochemistry* (Cambridge: Cambridge University Press, 1982).

28 D. Fitzgerald, "Beyond tractors: the history of technology in American agriculture", *Technology and Culture* 32 (1991): 114–26.

29 R.E. Kohler, *Partners in Science: Foundations and Natural Scientists* (Chicago: University of Chicago Press, 1991), p. 110.

30 Mowery & Rosenberg, *Technology*, pp. 161–8; and cf. with L. Busch & W.B. Lacy, *Science, Agriculture, and the Politics of Research* (Boulder: Westview Press, 1983).

31 R. Bud, *The Uses of Life: A History of Biotechnology* (Cambridge: Cambridge University Press, 1993), pp. 104–5; see also D. Hounshell & J.K. Smith, *Science and Corporate Strategy: DuPont R&D, 1902–1980* (Cambridge: Cambridge University Press, 1988), pp. 249–74.

32 Hounshell & Smith, *Science and Corporate Strategy*, p. 451.

33 *National Agricultural Chemicals Association Industry Profile* (Washington: Ernst & Young, 1992).
34 See P.A. Larkin & W.J.D. Stephen, *From Formalin to Fortran* (Ottawa: Science Council of Canada, 1971).
35 D. Verney, *Three Civilizations, Two Cultures, One Nation* (Durham: Duke University Press, 1986), p. 30.
36 See J. Ben David, *The Scientist's Role in Society: A Comparative Study* (Chicago: University of Chicago Press, 1971).
37 B.A. Kimmelman, "A Progressive era discipline: genetics and American agricultural colleges and experiment stations, 1890–1920", (Ph.D. dissertation, University of Pennsylvania, 1987); D.K. Fitzgerald, *The Business of Breeding: Hybrid Corn in Illinois, 1890–1940* (Ithaca: Cornell University, 1990); and J.R. Kloppenburg, *First the Seed: The Political Economy of Plant Biotechnology, 1492–2000* (Cambridge: Cambridge University Press, 1988).
38 T.H. Anstey, *One Hundred Harvests: Research Branch, Agriculture Canada, 1886–1986* (Ottawa: Research Branch, Agriculture Canada, 1986).
39 R.P. McIntosh, *The Background to Ecology. Concept and Theory* (Cambridge: Cambridge University Press, 1985).
40 S. Kingsland, *Modeling Nature. Episodes in the History of Population Ecology* (Chicago: University of Chicago Press, 1985).
41 See R.C. Tobey, *Saving the Prairies: The Life Cycle of the Founding School of American Plant Ecology, 1895–1955* (Berkeley: University of California Press, 1981). On genetics, see P. Palladino, "Wizards and devotees: the Mendelian theory of inheritance and the professionalization of agricultural science in Britain and the United States, 1900–1920", *History of Science* 32 (1994): 409–44.
42 Tobey, *Saving the Prairies*, p. 25; see also M. Nicholson, "National styles, divergent classifications: a case study from the history of American and French plant ecology" in L. Hargens *et al.* (eds.), *Knowledge and Society: Studies in the Sociology of Science, Past and Present* (Greenwich: JAI Press, 1989); pp. 139–86.
43 J. Harwood, "National styles in science: genetics in Germany and the United States between the world wars", *Isis* 78 (1987): 390–414.
44 R.W. Winks, *The Relevance of Canadian History: US and Imperial Perspectives* (Toronto: MacMillan, 1979).
45 R. Sawyer, "To make a spotless orange: biological control in California", (Ph.D. dissertation, University of Wisconsin, 1990).
46 J. Hagen, *An Entangled Bank: The Origins of Ecosystem Ecology* (New Brunswick: Rutgers University Press, 1992).
47 A. Pickering, *Constructing Quarks: A Sociological History of Particle Physics* (Edinburgh: Edinburgh University Press, 1984).
48 B. Latour, *Science in Action* (Cambridge: Harvard University Press, 1987).

CHAPTER 1

Entomology in the United States

ENTOMOLOGY IN THE "STONE AGE OF SCIENCE"?

Rachel Carson wrote in *Silent Spring* that "the concepts and practices of applied entomology for the most part date from the Stone Age of science".[1] In other words, the crisis which she had so famously and dramatically described was the direct consequence of entomologists' obsession during the 1940s and '50s with the development of ever more powerful and extremely ecologically disruptive insecticides. Some entomologists agreed with Carson's assessment, believing that, while searching for more effective insecticides, most of their colleagues had forgotten that the most enduring solutions to problems of crop protection had always been based on a sound knowledge of insect ecology. It certainly is odd that this amnesia was so total that, even though entomologists could boast of many past contributions to the intellectual development of ecology, by the mid 1950s they took the chemist rather than the ecologist or biologist as their paragon of professional excellence. In this chapter, I will argue that throughout the history of their discipline, most American entomologists were keenly aware of the ecological dimensions of their science, but a variety of institutional factors, some of which they themselves helped to shape, led them gradually to adopt a research agenda that ultimately undermined their identity as students of living organisms.

THE BEGINNINGS OF AMERICAN ENTOMOLOGY

During the first half of the nineteenth century, the interests of American naturalists did not differ very greatly from those of their transatlantic counterparts, toward whom they looked for intellectual leadership. Theological interests inspired them to collect the many unknown insect species which they were finding as the frontier of settlement moved westward, beyond the Appalachians. Many specimens were then sent for identification to knowledgeable metropolitan collectors, such as the members of the Entomological Society of Pennsylvania in

Philadelphia.[2] Lacking the extensive reference collections needed to settle the more difficult cases, these latter collectors often corresponded with their better endowed transatlantic counterparts, sharing information about the new species and discussing matters of systematics.[3] It was thanks to this network that Thomas Say, author of the first insect catalog prepared by an American, and Superintendent of Literature, Science, and Education in Robert Owen's New Harmony, was elected to the Linnaean and the Zoological Societies of London.[4] By the 1840s and '50s, naturalists like Say were both conscious of a distinct identity as "entomologists", and sufficiently numerous to organize short lived entomological associations, like the Entomological Society of Pennsylvania. The personal fortunes of the leading naturalists, which helped to establish these associations, were not, however, sufficient to create the reference collections needed to establish both their intellectual independence from transatlantic authorities, and their control over the activities of the increasing numbers of more amateur collectors across their rapidly expanding country.[5] The even less affluent among these naturalists were driven to exercise their entomological skills while employed as instructors in the expanding American system of higher education, which by the 1850s was committed to providing its students with a sound grounding in natural history and theology. However, the financial rewards of these situations often were very meager, and many of these instructors had to supplement their incomes by undertaking pursuits of a far, far more practical nature.[6]

Around the middle of the nineteenth century, American agriculture was being profoundly transformed. Growing urban populations at home, and more accessible markets abroad, increased the demand for its products. The expanding network of canals and railways, which greatly facilitated access to these domestic and foreign markets, also encouraged the capitalization and commercialization of grain farming on the abundant and fertile land west of the Appalachians.[7] Furthermore, the greater profitability of both grain production on the western frontier, and of cotton in the southwestern United States then led farmers on the eastern seaboard to abandon cultivation of these two crops for the still more profitable, but far more capital intensive production of fruits and vegetables. The resulting establishment of extensive monocultures in the West and Southwest, and of highly capital intensive production in the East exacerbated farmers' problems with insects. Extensive monocultures of alien crops provided hospitable environments for both native and introduced noxious species of insects, and capital intensive production led to greater concern over any damage these insects might cause. Technological innovations, however, had already helped American farmers to surmount major economic problems confronting them.[8] An acute shortage of skilled labor on the extensive and very fertile western farmland had encouraged the search for methods of improve labor productivity, and the development of mechanical equipment was one answer. Meanwhile, fertilizers had allowed farmers in the northeastern and the mid-Atlantic states to compete with their more productive western counterparts.[9] Entomologists could perhaps help them all with their latest problem. Thus, in 1854 the New York Legislature appointed Asa Fitch to the office of State Entomologist, the first such office in the Union.

In his first report to the New York State Agricultural Society, Fitch discussed the various ways to control the Plum Curculio, which was then damaging many local orchards.[10] Although most of his advice on protective measures only gave a stamp of scientific legitimacy to practices empirically derived by the affected growers themselves, he was also able to indicate which of these practices were most effective. For example, he could tell the growers that washing trees with tobacco infusions was useless, while running poultry in the orchards might be of some aid against the Curculio. These growers' willingness to support Fitch, and many other entomologists beside him, however, was not just economic. These entomologists' advice was based on studies of the life cycles of plants and insects as they displayed themselves in the orchard and farm. Like farmers, they lived and learned in the field. Moreover, studying insects was for both of them one of the many possible ways to understand a divinely established order of nature, at a time of intense debate about Charles Darwin's theory of organic evolution, and one of the key elements of this natural theology was the notion of a "balance of nature" checking the reproductive capacities of all plants and animals, including insects.[11] Entomologists and other agricultural experts often referred to it to explain the origin the farmer's economic problems. In 1858, for example, the editor of the *Prairie Farmer* wrote that,

> birds, reptiles, and predacious insects were doubtless provided to keep the most injurious species in check ... But man interfered, and reversed the natural order ... The forests which fed the destructive insect and sheltered the devouring bird and reptile, have been swept away, and the countless billions of the one and the frightened remnants of the other, seek food where it can be found, or safety in less 'civilized' regions; and the insect has now settled in our gardens and fields, and his enemies, and our friends are fast disappearing before the tide of 'improvement', and ... ignorant prejudice or wanton barbarity.[12]

Such explanations were much more familiar to farmers than the much more arcane ones about the action of fertilizers, forged by agricultural chemists in their mysterious laboratories.

Farmers' support for entomology was, nonetheless, limited; between the 1850s and '70s, only Illinois and Missouri followed the example set by the New York legislature.[13] Furthermore, it had its price: when the New York legislature appointed Fitch to his new post, the New York State Agricultural Society criticized earlier entomological investigations conducted under the aegis of the New York Natural History Survey for excessive emphasis on systematics. It argued that the surveyors' final report was,

> too purely scientific and [that] it would be of little consequence to know that a particular kind of moth or fly is an inhabitant of this state, unless [the society was] also informed of its history and habits, and whether it is a depredator upon any substance which is of value to man.[14]

Since expert knowledge of insect systematics had served to set entomologists apart from more amateur collectors, de-emphasizing this aspect of their endeavors threatened to undermine their institutional status.[15] To maintain it, they began to abandon work on taxonomy and systematics, and to focus their attention, as Fitch did, much more determinedly on the life histories of economically important insect species; work on the natural history of insects had now to be made into a new, scientifically rigorous endeavor, the study of insect "ecology".[16] This first transformation of entomology was greatly facilitated by the emergence in the 1880s of a national organization for the agricultural research.

AGRICULTURAL EDUCATION AND RESEARCH

During the 1860s, there were repeated efforts to establish a national organization to support agricultural education. In 1862, for example, the United States Congress voted into law the Morrill Act. Each state of the Union was granted ownership over federal lands, whose sale was then to be used to fund the establishment of a college for teaching "without excluding other scientific or classical studies … such branches of learning as are related to agriculture and the mechanic arts … in order to promote the liberal and practical education of the industrial classes".[17] Despite the implicit ambitions to create a gentile yeomanry, the establishment of these colleges was little more than a political gambit to strengthen the then critical alliance between the farming western states and the industrial eastern states — the country was then in the midst of the Civil War. In the East the endowment funds went to existing institutions for higher learning, and in the West there were too few students and teachers in the new, land-grant colleges to have any impact on the agricultural community. A year later, Congress also established the United States Department of Agriculture to better serve the needs of American farmers by providing them with the services of "chemists, botanists, entomologists, and other persons skilled in the natural sciences pertaining to agriculture".[18] The Civil War, however, was about the relative authority of federal and state governments. It is not surprising, therefore, that many legislators were deeply concerned that the newest branch of the federal government might attempt to interfere in the economic affairs of the various states as it did with slavery. They sought, therefore, to hand it as little authority possible to change any aspect of American agriculture. Thus, for example, the proposal by Townend Glover, the first Federal Entomologist, to reduce the problems caused by introduced pests by establishing import controls and quarantines was blocked as unacceptable federal interference in matters of commerce.[19] Federal involvement in agricultural education and research came to a halt after these first initiatives.

A completely different political agenda for agricultural education and research began to emerge, however, in the late 1870s and early 1880s. After the passage of the Homestead Act, the Western frontier moved rapidly into the Great Plains, and wheat cultivation came to dominate in this area of the country. Since the organization of marketing was still weakly articulated, this very rapid expansion of wheat production resulted in large surpluses; wheat prices collapsed and a severe agricultural depression ensued. The situation was aggravated by adverse natural factors, among which was a series of epidemic outbreaks of the devastating Western Grasshopper.[20] This combination of economic and natural calamities marked the beginning of the "Agrarian Revolt", which was profoundly influential in reshaping the American political landscape of the late nineteenth century.[21]

Despite continuing, strong opposition to any federal intervention in the world of commerce, Congress adopted a number of different palliative measures against this agrarian crisis. Among them was the establishment of the three man United States Entomological Commission to deal with the Grasshopper, which was placed interestingly under the aegis of the Department of the Interior rather than the Department of Agriculture. Charles Valentine Riley, who as the State Entomologist for Missouri had masterminded its organization, was selected to head the Commission. The other two members were Cyrus Thomas and Alphaeus Packard. These three entomologists' subsequent studies of the Grasshoppers' life cycles, and of the impact of climatic conditions on its abundance, provided some useful information for western farmers: it was now possible to predict the timing, extent, and probable severity of outbreaks, which might allow farmers to adjust their planting schedules so as to minimize the impact of the outbreaks.[22] This work catapulted the Commission into the public limelight; Riley succeeded in putting himself and entomology in a very strong position in the national scientific arena. In 1878 he was appointed Federal Entomologist. Exemplifying all the problems of political patronage and corruption that incensed late nineteenth century "progressive" reformers, Riley then used his new position to both transfer the powers of the Entomological Commission to entomologists in the United States Department of Agriculture, and form the Bureau of Entomology, which he headed until 1894.

The increasing acceptance of a limited role for government in solving social and economic problems was also evident in the land grant colleges. During the 1880s, a new generation of administrators and scientists came to believe that their institutions should be more closely tied to agriculture, perhaps by taking more seriously Morrill's call for "a careful, exact, and systematized registration of experiments". A number of agricultural research institutions had already been organized in this way by various legislatures in the wealthier eastern seaboard states. Since fertilizer fraud was a major problem confronting farmers in the region, the major task for the scientists employed by these stations was to analyze fertilizers and assess their utility to farmers.[23] Thus, what little research took place in these experiment stations focused on the chemistry of fertilizers. This situation was not exactly appreciated by the scientists in the stations, for they aspired to a more creative role. According to them, mere checking on the quality of existing fertilizers was a waste of their unique preparation. There should instead be a legislative framework prohibiting the sale of ineffectual

Prior to the emergence of arsenical insecticides, entomologists and farmers alike had believed that the modification of cultural practices, to make crops less attractive havens for the insect species attacking them, was the most effective pest control strategy.[32] Many of these so-called "cultural control" practices were often developed by the farmers themselves in a very empirical manner, by trial and error, and by relying on some knowledge of the habits of any noxious insect species. But, when they faced the problems caused by the sudden appearance of newly introduced species, they had to turn to the scientists in the experiment stations and the United States Department of Agriculture for a quick solution. For example, in the 1890s the Cotton Boll Weevil was devastating the cotton industry, and the United States Department of Agriculture's entomologists were called to respond to this crisis. Their investigations soon suggested that if cotton growers were to harvest only the earliest bolls, and were to remove the plant stalks by plowing over the field at the end of the season, the Boll Weevil would have no over-wintering refuge and its numbers during the following season would be smaller.[33] Unfortunately, the land tenure system prevailing in the cotton growing regions made these particular cultural modifications, sometimes known as the "government method", highly unpopular among everyone but plantation owners. The majority of growers, small tenant farmers and sharecroppers, enjoying uncertain rights to tenancy, wanted all that the land they rented would yield, and hence were not interested in collecting only the early cotton bolls and destroying the crop before it had set its last fruit. The soundest advice that these federal entomologists could give the grower, on the basis of their ecological approach, was unacceptable within the existing social organization of cotton production (or wherever else producers were small tenant farmers and sharecroppers).[34]

An alternative preventive strategy was pioneered in 1880s, when the Cottony Cushion scale was accidentally introduced in California and threatened to destroy the nascent, but already well organized, citrus industry.[35] Responding to this problem, entomologists in the United States Department of Agriculture proposed to establish the point of origin of the Scale, then to identify its local natural enemies, and finally to introduce those enemies into California citrus groves. In 1888 they successfully accomplished all this; they introduced the Vedalia Beetle and the Scale problem promptly disappeared. This approach to crop protection, the "biological control" approach, immediately caught the imagination of many entomologists, who sought to use it to solve otherwise intractable problems such as those posed by the Boll Weevil, the Codling Moth or the Brown Tail Moth. However, this approach required extensive work to identify precisely the insect species damaging the crop and then further and equally extensive work to locate the appropriate natural enemies. Often, this final step also required the willingness of administrators to support travel abroad in search of these enemies. Unfortunately, when this was all done, the effort failed most times to yield the spectacular success obtained with the Cottony Cushion Scale.

Thus, once highly effective, arsenical insecticides appeared on the market in the 1870s, their use became the most attractive strategy for crop protection. For example, apple orchards were often devastated by the Codling Moth. Cultural practices to control this pest were unacceptable as they required much labor and sacrificed productivity by requiring, for example, the removal of older trees, spacing them further apart, and maintaining clean orchards. The more interventionist introduction of natural enemies did not yield anything like the spectacular demise of the Cottony Cushion Scale. In contrast to this, the effects of treatment with an arsenical insecticide, Paris Green in this case, were seen quite readily, as all the grower had to do was count the number of dead insects on the ground.[36] Moreover, unlike cultural control and biological control, chemical control did not require any coordination of neighboring farmers' activities, as each of them could now protect himself or herself regardless of what the neighbors were doing. Lastly, since farmers turned to the scientists in the experiment station only in a time of crisis, demanding immediate help, all that the latter could do was to recommend the use of some insecticide. To do otherwise would have undermined their institutional security, for farmers would have been most unlikely to continue funding or otherwise supporting any entomologists who persisted in providing solutions to their problems which were slow, impracticable, or ineffectual.

ENTOMOLOGISTS, INSECTICIDES, AND ENTOMOLOGICAL RESEARCH

The emergence of insecticides like Paris Green and London Purple began to change the attitude of many entomologists toward them, for these were no quack's useless potions but very effective remedies. Demarcations between preventive and remedial measures were eroded and became less distinct. In the past, for those entomologists who viewed agricultural production primarily as a scientific, ecological problem, prevention had meant altering the biological environment in which the crops were grown so as to make them less attractive to insects feeding on them. However, as the land-grant colleges produced increasing numbers of graduates in entomology, many of whom were the children of farmers, the average entomologist tended to be more attuned to the views of farmers, more sensitive to their needs and priorities.[37] For them, prevention now came to mean simply destroying insect populations before they could do any damage.

The focus of entomological work was now on preventing economic loss rather than preventing the growth of insect populations. The United States Department of Agriculture suggested that insecticidal treatment was a good preventive measure, which should be adopted by growers as an insurance policy; it always paid to spray regardless of whether the Codling Moth, for example, was or was not inflicting any damage at the time.[38] Furthermore, insecticides came to be viewed by many entomologists and farmers as one of the hallmarks of progress in agriculture. In 1903 the *United States Department of Agriculture Farmers' Bulletin* suggested that "the more progressive and business-like apple growers are the staunchest advocates of spraying, and their efforts are uniformly successful ... spraying always brings about greater efficiency and reduction of expenses".[39] For entomologists, investigations in the laboratory on chemical insecticides seemed novel, exciting. Thus, in 1908, Henry Fernald from the State College of Massachusetts wrote that,

> the use of arsenic poisons in the control of insect pests ... is one of the fundamental principles of economic entomology. Yet, a careful examination of our actual knowledge of arsenic in their relation to insect and plant life gives surprising results, showing how little is really known and how much is merely empirical, and indicates that a broad field for chemical, entomological and physiological research is waiting for explorers.[40]

Entomology, thanks to these very effective insecticides was becoming both a truly economic and fully professionalized endeavor.[41]

Not all entomologists were, however, satisfied with the rising popularity of chemical insecticides. They all readily acknowledged that the deadly effects of chemical insecticides were more persuasive than the less immediately visible benefits of cultural modifications and introductions of natural enemies, and that farmers were for this reason strongly inclined to use them. Yet, the varying quality and effectiveness of chemical insecticides sold in an unregulated market often caused farmers and growers to seriously damage their plants.[42] To protect the growers, and bolster their own professional position, entomologists in the United States Department of Agriculture suggested to them that they send samples of these insecticides to the state agricultural experiment stations for testing. This saddled entomologists in the laboratories of these stations with more of those tedious chemical analyses which had much burdened agricultural chemists before them. More significantly, influential entomologists such as Herbert Osborn at Ohio State University, complained that very little knowledge of insect biology beyond feeding habits was needed for this work and to improve the effectiveness of the insecticides.[43] But, as entomologists in the federal Bureau of Entomology showed, insects could escape the deadly action of Paris Green through habits other than those related to feeding. More specifically, cotton growers were using Paris Green early in the season hoping to eliminate immature Boll Weevils, but the practice, these entomologists discovered, was utterly useless because at this stage the insect was deeply ensconced within the cotton fruit and was thus well protected from the poison.[44] Not surprisingly, therefore, the various authors of the papers appearing in the *United States Department of Agriculture Farmers' Bulletin* continually stressed that developing an effective insecticide required keen insight into the life history of insects. But Osborn and others feared that this knowledge was becoming rarer because entomologists depended so heavily on farmers and growers that they had to keep their nose to the grindstone, testing (and perhaps developing) more and more insecticides. It would take a very strongly willed director to resist the pressure to continue this tradition and to encourage instead studies of natural history to improve the effectiveness of existing insecticides. Entomologists were now hired to stay in the laboratory, not to play the field naturalist.

By the early years of this century, however, some entomologists began to complain vociferously that they had to spend so much time investigating the merits of the various insecticides that it was not possible to conduct any serious investigations at all while in the service of the state agricultural research establishment. For example, leading entomologists, such as Mark Slingerland, a former student of John Henry Comstock at Cornell University, complained that,

> with such inspection or police duties [i.e.: controlling the quality of insecticides appearing on the market], more or less teaching, an ever increasing correspondence, farmers' institute work, and other necessary duties, it is a wonder that many of the entomologists find any time for investigational work, and yet much good work is done every year. But careful survey of the published results reveals the fact that more original work is coming from those who have little college or station work. This result is inevitable, for good investigational work requires constant observation, attention to details, and unlimited, unhampered time. No day is long enough for a man to inspect nurseries, attend farmers' institutes, teach several hours, write five to a dozen letters, eat and sleep, and then attempt a careful study of some destructive insect. Some of this work, which no entomologist ought to shirk, must be left half done, and unfortunately it is too often that the original investigation suffers.[45]

Slingerland sought to relieve this situation of American entomologists by pressing for a reform of the institutional structure supporting their work. The Adams Act of 1905 provided additional support in the state agricultural experiment stations for "original" research, drawing a distinction between original and repetitive. According to Slingerland, this act,

> provides a way for more abstract research in general and fundamental principles covering the ecology of insect pests and their control. Valuable results should come from an exhaustive study of the factors — climatic, organic, etc. — which influence the health, or what I call the ups and downs or periods of great destructiveness and comparative scarcity of insect pests. Some entomologists hope to see the time when they can predict or state the probabilities of excessive depredations by certain insects with as much success as the weather forecasters sometimes predict. Owing largely to the effective work of parasitic enemies we can already predict with considerable certainty that the most destructive periods of the Hessian fly and army worm will not continue for more than a year or two in the same locality.[46]

If all that could be expected of this "more abstract" ecological research was the assurance that the devastation of wheat fields due to an epidemic of the Hessian Fly would subside within a year, it is no surprise that many growers and legislators continued to be quite skeptical about the value of such original investigations. In fact, not all entomologists agreed with Slingerland's views on the Adams Act. For example, Robert Forbes from the state agricultural experiment station for Arizona rejected them stating that,

> if there is one sentence that expresses our ... attitude toward our work and toward the public, it is this, that it is not science for science's sake alone; it is certainly not science for personal reputation's sake; but it is science for humanity's sake that we are engaged in pursuing.[47]

It would appear, however, that with the implementation of the Adams Act those entomologists who, like Slingerland, viewed their chief responsibility as research, rather than policing and advisory work, won the day. On the other hand, during the next decade, they were forced to seek still more legislative action to promote their vision of entomological work. For example, three years after the passage of the Adams Act, the American Association of Economic Entomologists pressed Congress for passage of legislation regulating the sale of insecticides, hoping thereby to conclusively eliminate any policing burdens.[48] However, it was not until the passage of the Smith-Lever Act of 1914, which was intended to improve the links between the experiment stations and the growers through the creation of the Extension Service, that these entomologists were finally able to deflect onto extension workers any direct pressure from the local agricultural community.[49] By the 1920s, the president of the American Association of Economic Entomologists, Arthur Ruggles, could argue that, in the past, the practical responsibilities of economic entomologists had led them to think more in terms of remedial solutions. They had not been allowed enought time to understand exactly how insects damaged their clients' crops, and what might be done to prevent this over a longer term than one day to the next. Now, the situation was changing. So much had been accomplished in recent years to assure entomologist some independence from their local patrons. The more desirable ecological perspective on the problems of economic entomology was finally gaining hold. It was Ruggles' impression that,

> the use of insecticides has led to a side issue that has assumed enormous proportions. Many ... entomologists even before thoroughly studying the life of the insect [begin] experimenting with insecticides for the control of the insect in question. An enormous field has opened up here for the entomologist but it must of necessity be primarily a chemistry field. It is not well for the entomologist to lose sight of the insect. The insect is primary not secondary. Not all insects can be controlled by insecticides and the sooner we learn that fact the better ... While this phase [has] been going on in the side line of chemical development other entomologists [have] already begun the control study from another angle ... This is a control based on the ecological study of insects in the broadest sense of the term.[50]

However, in 1939, this now elder statesman of American entomology complained bitterly about his own lack of freedom to study of the ecology of grasshoppers, in spite of all the years he had spent aiding growers in his home state of Minnesota.[51] Despite all the distancing between scientists and their agricultural constituency, some scientists still felt that their situation had changed very little between the 1880s and the 1930s.

EDUCATIONAL REFORM: CHEMISTRY OR ECOLOGY?

During the 1920s, Arthur Ruggles and many of his colleagues had hoped that the newer generation of entomologists would bring about some fundamental change to the orientation of their science. They hoped that these new entomologists would be endowed with "a vision beyond the things we are doing in economic entomology today and see what a tremendous realm our favorite science occupies in the affairs of men, offering wonderful opportunities for study, recreation and thought".[52] Unfortunately, for reasons having to do with their educational backgrounds, the newer generation of entomologists did not fulfill these hopes.

In 1899, John Smith, the chairman of the Entomology Section of the American Association of Agricultural Colleges and Experiment Stations, had explained that there were two types of academic institutions teaching entomology. One group aimed to provide a very broad education, enabling the students to pursue any type of career related to the life sciences. Institutions like John Henry Comstock's Cornell University or William Morton Wheeler's Harvard University fell into this category.[53] The second group, which included most of the land-grant colleges, aimed mainly to prepare the students for more practical undertakings. Smith described the goals of the institutions of this second class as simply to prepare graduates for economic work or advanced research in the better equipped facilities of the more prestigious universities.[54] By the time of Ruggles' address to the American Association of Economic Entomologists, this situation had become a source of major concern among leading entomologists. There was a concerted effort to reform instruction in entomology. But with what subjects should the future entomologist be conversant?

Given the rising popularity of chemical insecticides among farmers, especially the more progressive farmers, the ultimate judges of the worth of any crop protection strategy, it is not surprising that entomologists might want to emphasize instruction in chemistry. In fact, in 1923, an entomologist at the University of Minnesota, William Moore, wrote in the *Journal of Economic Entomology* that

> the economic entomologist justifies his work and makes his claim for federal and state funds upon his ability to make sound recommendations for the control of various insect pests. Altho [sic] parasites may hold in check certain insects and damage from others may be reduced by proper cultural methods, the burden of control rests upon the use of insecticides. Progress in the development of insecticides and a knowledge of their action has not kept pace with the development of other phases of economic entomology. The failure of the entomologist to make greater progress in the development and use of insecticides is no doubt due to his lack of chemical knowledge. The chemical training of the entomologist in the past has been so meager that he makes and recommends mixtures of insecticides and fungicides which are not only incompatible but even useless and in many cases harmful to the plant.

In other words, to provide more effective service, economic entomologists had to pay more attention to chemical rather than biological control, and have a sounder preparation in chemistry (as well as physiology and toxicology).[55]

Genetics and evolutionary biology do not seem to have been considered equally worthy subjects for students' attention. It is then not surprising that many entomologists were at a loss to explain why the usual sulfured lime treatments used to control the San José Scale appeared to become increasingly ineffectual during the first decade of this century. Even less surprising is that, when an explanation was offered in 1913, by Axel Melander, it fell on deaf ears.[56] Having obtained what was commonly viewed as one of the finest possible educations in entomology, under Wheeler at Harvard University, Melander turned to a simple theory of population genetics to suggest that these treatments had resulted in the selection of a strain that was not affected by the sulfured lime, and that it was very fortunate that the available insecticides were not as powerful as some entomologists would have liked, for, otherwise, their use would have resulted in the evolution of a far greater number of resistant insect species. He closed his argument by making "the strange assertion that the more faulty the spraying this year the easier it will be to control the scale the next year".[57] Few entomologists had the educational grounding to understand this message.[58] Immediate, and ever more effective answers were what counted, and work in chemistry, insect physiology, and toxicology would continue to attract more funding and students.

THE SEARCH FOR EXTERNAL SUPPORT

It is important to realize that the shortage of funds for research in the experiment stations was general; it did not really matter whether the research was aimed at obtaining greater insight into the chemistry of insecticides, or into the ecology of insects. This forced investigators to seek external sources of support. The ensuing, widespread efforts to capture the attention of external supporters gradually led to a polarization of the entomological profession along the lines of chemical versus biological control. In his 1924 presidential address to the American Association of Economic Entomologists, Arthur Ruggles had pleaded for moderation with the following words,

> in every great movement whether political, religious or scientific, ideas seem to swing from one extreme to the other. For the general run of people there seems to be no middle ground. They must be either extremely radical or extremely conservative. My idea at this time is that we as scientists should not be too impulsive. We should be deliberate and not change radically until all the scientific facts are brought out and marshaled in evidence. We should not be too hasty to change nor so conservative that we cannot change when the truth comes.[59]

But such moderation was difficult in the heat of competition for external support. For example, in 1935, Robert Vickery of the California Spray Chemical Corporation wrote in the *Journal of Economic Entomology* about the concerns of the chemical insecticide industry over the impact of the policies proposed by president Franklin Delano Roosevelt's secretary of agriculture, Henry Wallace, to solve the agricultural problems of the Great Depression. Wallace was planning to take land out of production to reduce the glut of agricultural products on the national and world market, but, as Vickery pointed out to the readers of the *Journal of Economic Entomology*, overproduction was a problem of the agricultural industry as a whole, and the economic survival of the individual farmer or grower during this difficult period rested on more efficient production. The problem for economic entomologists and the insecticide industry was then to foresee how these producers would increase their efficiency. Would it result, Vickery asked, "in increased consumption of insecticides? ... Will the trend in economic entomology be toward chemical control or toward biological control?"[60] He answered these questions by arguing that over the past decade the cost of insecticides had constantly declined, and as this pattern continued, they could be applied in a wider variety of situations. Many situations, where the use of biological controls had previously proved more economical, could now be managed with cheaper and more effective chemical insecticides. The real problem for economic entomologists, then, was to accelerate the development of even more effective and cheaper insecticides. This was helped along in 1936, by the establishment by the chemical insecticide industry of the Crop Protection Institute, which would focus on fundamental studies relevant to the development of new and more effective insecticides.[61]

The Crop Protection Institute was just the most visible result of entomologists' increasing ties to the chemical insecticide industry. Support was much more likely to be given in smaller forms, and everyone in the experiment stations was encouraged to seek it. This, at first, greatly concerned the influential Association of Land-Grant Colleges and Universities; in 1928, it issued guidelines on the management of external grants to the personnel of the experiment stations, aiming to insure that they would not require the scientist to undertake research that was too narrowly directed by the priorities of the donor. Such requirements would have been flagrant violations of the stipulation that scientific research in the experiment stations should benefit the general public.[62] Such external support, however, was, for the directors of the experiment stations, the heads of college departments, and individual researchers in these institutions, quite irresistible. It provided everyone with some shelter from pressure by their respective superiors and other agricultural constituencies to economize or work on more economically significant projects. For example, since the Legislature of the State of Minnesota was not interested in providing the Division of Economic Zoology at the University of Minnesota with funds for the purchase of an exhaust hood for Harold Shepard's toxicological investigations, Shepard obtained them from the longtime corporate ally of the Division, the American Cyanamid Company.[63] This hood allowed him to study, among other things, the effects of American Cyanamid fumigants. Furthermore, in 1936 he entered into an arrangement with the Röhm & Haas Corporation which allowed him to continue his usual research on insecticides in full freedom, assisted by a graduate student supported through a grant from the corporation. The only stipulation was that the research should include the study of one particular chemical compound of interest to Röhm & Haas.[64] The insecticide industry was well served by these arrangements because they shifted onto the taxpayer the burden of capital investment on facilities and personnel, which would have otherwise had to be carried by the industry itself.

Entomologists interested in toxicology, physiology, or other subjects closely connected to the development of more effective chemical insecticides, were not the only beneficiaries of these financial arrangements. Sometimes, even ecological research could be supported through grants from the chemical industry and other corporate donors. For example, in the 1920s and '30s, Royal Norton Chapman, the chairman of the department of entomology at the University of Minnesota, and one of the foremost population ecologists of the period, used generous grants from the ISCO Corporation to support his assistants in testing the effectiveness of chloropicrin against the Flour Beetle.[65] In so doing, Chapman also fulfilled his duties to two of the station's major constituents, General Mills and Quaker Oats, who he eventually advised to use ISCO chloropicrin. This freed up federal funds to study the population dynamics of *Tribolium confusum*, the more academically respectable name of the humble Flour Beetle, easily leaving the impression that these investigations were immediately relevant to the interest of millers and insecticide manufacturers.[66]

Everyone seemed to gain from this alliance of economic entomology and the insecticide industry. The one, most obvious consequence, however, was an increasing orientation of entomological research and practice toward the study and use of chemical insecticides.

ECONOMIC ENTOMOLOGY IN THE AGE OF DDT

By the middle of the 1930s, the use of lead and arsenic insecticides to protect apple crops was causing severe problems. Residues left on apples for the British market became a major issue with the United States Public Health Service and the Food and Drug Administration, because they were being used to exclude American produce from crucial export markets such as Britain and her Empire. Moreover, these insecticides were poisoning the trees, especially those in the most commercially oriented orchards of Washington.[67] The immediate answer to these problems would have to come from better use and formulations of lead and arsenic insecticides, but in the longer term what was needed was stepping up investments in the search for new insecticides.[68] The necessities of war boosted this search far more powerfully than anyone could have previously imagined. Responding to a shortage of botanical insecticides like pyrethrum and rotenone, which were usually imported from now inaccessible East Africa and Southeast Asia, the Office of Scientific Research and Development of the Department of War determined that the production on new insecticides should be a major goal of entomological research in the

plantation owners then turned to the much more toxic organophosphate insecticides, and again selected for resistant insect strains. In the meantime, all predators and parasites were completely annihilated, and any effect they might have had on the populations of insect species damaging the cotton crop vanished. In the face of mounting costs of control, the cotton industry collapsed. Only after an intensive campaign to ban the use of most insecticides was it possible to resuscitate cotton production.[89] In other words, entomologists were failing to meet their chief responsibility: increasing agricultural productivity. Many of them, especially those who were called to resolve these crises, argued ceaselessly that the used chemical insecticides had to rest on some understanding of its ecological and evolutionary implications. They stressed time and again that an ideal solution was to develop programs which caused the least possible damage to native or introduced predators and parasites. The broad spectrum action of synthetic organic insecticides, however, seemed to preclude any selective action against any one species alone, and the destruction of natural enemies seemed unavoidable. More significantly, entomologists had become so divided that any constructive interaction between the proponents of chemical and biological control seemed impossible to Ken Hagen, from the Division of Biological Control in the California state agricultural experiment station in Berkeley.[90] Much would have to change in the institutional organization of American economic entomology if this alternative, one which had been available at least since the 1930s, was to become more than just an interesting idea.

NOTES

1 R. Carson, *Silent Spring* (Boston: Houghton Mifflin, 1987), p. 297.

2 C. Sorensen, "Brethren of the net: American entomology, 1840–1880", (Ph.D. dissertation, University of California, Davis, 1984), pp. 1–109.

3 Sorensen, "Brethren", pp. 28–63. See also G.H. Daniels, *American Science in the Age of Jackson* (New York: Columbia University Press, 1968), pp. 6–33; and *Science in American Society. A Social History* (New York: A.A. Knopf, 1971), pp. 101–25.

4 Sorensen, "Brethren", p. 12.

5 Sorensen, "Brethren", pp. 45 and 49.

6 See S.G. Kohlsted, "Curiosities and cabinets: natural history museums and education on the antebellum campus", *Isis* 79 (1988): 405–26.

7 See W. Cronon, *Nature's Metropolis: Chicago and the Great West* (New York: Norton, 1991).

8 See M.W. Rossiter, "The organization of agricultural improvement in the United States, 1785–1865" in A. Oleson & S.C. Brown (eds), *The Pursuit of Knowledge in the Early American Republic* (Baltimore: Johns Hopkins University Press, 1976), pp. 279–98.

9 F.A. Shannon, *The Farmer's Last Frontier* (New York: Farrar & Rinehart, 1945), pp. 125–47; see also M.W. Rossiter, *The Emergence of Agricultural Science: Justus Liebig and the Americans, 1840–1880* (New Haven: Yale University Press, 1975).

10 Sorensen, "Brethren", pp. 124–6.

11 Sorensen, "Brethren", pp. 164–89. For a more general discussion of the history of the "balance of nature", see F.N. Egerton, "Changing concepts of the balance of nature", *Quarterly Review of Biology* 48 (1973): 322–50.

12 Quoted in Sorensen, "Brethren", p. 181.

13 Sorensen, "Brethren", pp. 111–63.

14 Quoted in Sorensen, "Brethren", pp. 123–4; see also p. 239.

15 Sorensen, "Brethren", p. 32.

16 For a parallel history of competition with amateur naturalists, in plant ecology, see R.C. Tobey, *Saving the Prairies: The Life Cycle of the Founding School of American Plant Ecology, 1895–1955* (Berkeley: University of California Press, 1981), p. 25.

17 Quoted in L.O. Howard, *A History of Applied Entomology* (Washingtons: Smithsonian Institution, 1930), p. 46.

18 Quoted in A.H. Dupree, *Science in the Federal Government* (Baltimore: Johns Hopkins University Press, 1986), p. 150.

19 Sorensen, "Brethren", pp. 234–5; and Howard, *History*, pp. 35–41.

20 Shannon, *Farmers*, pp. 291–328.

21 R. Hofstadter, *The Age of The Reform* (New York: Vintage Press, 1955), pp. 23–59.

22 Sorensen, "Brethren", pp. 223–73.

23 A.C. True, *A History of Agricultural Extension Work in the United States* (Washington: United States Government Printing Office, 1928), p. 107; see also J.M. Gregory as quoted in L. Busch, & W.B. Lacy, *Science, Agriculture, and the Politics of Research* (Boulder: Westview Press, 1983), p. 9.

24 For a review of contemporary economic and political developments, and their implications for the institutional organization of agricultural science in America, see B.A. Kimmelman, "A Progressive era discipline: genetics and American agricultural colleges and experiment stations, 1890–1920", (Ph.D. dissertation, University of Pennsylvania, 1987), pp. 21–78.

25 "Introductory", *Annals of the Entomological Society of America* 1 (1908): 1–2, on p. 1.

26 C.V. Riley, "Minutes of conference of experiment station workers", *Proceedings of the American Association of Agricultural Colleges and Experiment Stations*, 3 (1889): 105–6; and "The outlook of applied entomology", *Report of the Second Annual Meeting of the American Association of Economic Entomologists (1890)*, pp. 20 and 25.

27 See P. Henson, "Evolution and taxonomy: J.H. Comstock's research school in evolutionary entomology at Cornell University, 1874–1930", (Ph.D. dissertation, University of Maryland, 1990), pp. 405–6.

28 "Introductory", p. 2.
29 True, *Extension*, pp. 132–4.
30 See C.E. Rosenberg, *No Other Gods. On Science & American Social Thought* (Baltimore: Johns Hopkins University Press, 1976), pp. 173–84; and A.I. Marcus, *Agricultural Science and the Quest for Legitimacy: Farmers, Agricultural Colleges and Experiment Stations, 1870–1890* (Ames: Iowa State University Press, 1985), pp. 7–26.
31 Wheeler quoted in M.A. Evans & H.E. Evans, *William Morton Wheeler, Biologist* (Cambridge: Harvard University Press, 1970), p. 180.
32 C. Marlatt, *Important Insecticides*, United States Department of Agriculture Farmers' Bulletin No. 19, (1894), pp. 18–9.
33 L.O. Howard, *Insects Affecting the Cotton Plant*, United States Department of Agriculture Farmers' Bulletin No. 47, (1897), pp. 16–23.
34 F. W. Mally, *The Mexican Cotton-Boll Weevil*, United States Department of Agriculture Farmers' Bulletin No. 130, (1901), p. 23.
35 The history of this strategy in California is discussed in great detail by Richard Sawyer in "To make a spotless orange: biological control in California", (Ph.D. dissertation, University of Wisconsin, 1990).
36 Paris Green was the first in a long line of increasingly effective arsenical compounds, developed from late 1860s onwards; see E.O. Essig, *A History of Entomology* (New York: MacMillan, 1931), pp. 423–8.
37 On the education of agricultural scientists during this period, see M.W. Rossiter, "Graduate work in the agricultural science", *Agricultural History* 60 (1986): 37–7.
38 H.P. Gould, *Practical Suggestions for Fruit Growers*, United States Department of Agriculture Farmers' Bulletin No. 161, (1902), p. 15.
39 C.B. Simpson, *The Control of the Codling Moth*, United States Department of Agriculture Farmers' Bulletin No. 171, (1903), p. 13.
40 H. Fernald, "The fundamental principles of spraying", *Journal of Economic Entomology* 1 (1908): 265–267, on pp. 265–6.
41 See A.D. Rodgers, *Liberty Hyde Bailey: A Story of American Plant Sciences* (Princeton: Princeton University Press, 1949), pp. 71 and 135.
42 C.V. Piper, *Orchard Enemies in the Pacific Northwest*, United States Department of Agriculture Farmers' Bulletin No. 153, (1902), p. 23.
43 H.F. Osborn, "Entomological work in the experiment stations", *Proceedings of the American Association of Agricultural Colleges and Experiment Stations* 8 (1894): 85–6.
44 W.D. Hunter, *The Use of Paris Green in Controlling the Cotton Boll-Weevil*, United States Department of Agriculture Farmers' Bulletin No. 211, (1904), 21–2.
45 M. Slingerland, "The more urgent problems in insect control", *Proceedings of the American Association of Agricultural Colleges and Experiment Stations* 21 (1907): 104–10, on p. 105. See also Rosenberg, *No Other Gods*, pp. 173–84.

46 Slingerland, "Problems", p. 108.
47 R.H. Forbes in reply to Slingerland, on p. 105.
48 E.D. Sanderson, "Statement by the standing committee on proprietary insecticides", *Journal of Economic Entomology* 1 (1908): 159.
49 True, *Extension*, pp. 100–15.
50 A.G. Ruggles, "Pioneering in economic entomology", *Journal of Economic Entomology* 17 (1924): 34–41 on pp. 39–40.
51 University of Minnesota Archives: Division of Entomology and Economic Zoology: papers, 1890–1959: group ALD9E.1: folder 401.
52 Ruggles, "Pioneering", p. 35.
53 See Osborn, *Fragments*, pp. 93–120; and Howard, *History*, pp. 70–8.
54 J.B. Smith, "Report of the section on entomology", *Proceedings of the American Association of Agricultural Colleges and Experiment Stations* 12 (1898): 24–7, on p. 26.
55 W. Moore, "The need of chemistry for the student of entomology", *Journal of Economic Entomology* 16 (1923): 172–6, on p. 172–3. See also V.I. Safro, "Extension entomology in relation to research", *Journal of Economic Entomology* 22 (1929): 553–6.
56 A.L. Melander, "Can insects become resistant to sprays?", *Journal of Economic Entomology* 7 (1914): 167–72.
57 Melander, "Can insects become resistant to sprays?", p. 171.
58 This disregard for the evolutionary aspects of entomology can be viewed also as a symptom of the uncertain scientific status of evolutionary theory, and its politically embattled situation in the American educational curriculum (especially in rural areas); see E.J. Larson, "Before the crusade: evolution in American secondary education before 1920", *Journal of the History of Biology*, 20 (1987): 89–114.
59 Ruggles, "Pioneering", pp. 34–5.
60 R.K. Vickery, "Pest control and the New Deal", *Journal of Economic Entomology* 28 (1935): 1078–9, on p. 1078.
61 See W.C. O'Kane, "The Crop Protection Institute", *Journal of Economic Entomology* 29 (1936): 6–20; and D.M. DeLong, "Industry's challenge to young entomologists", *Journal of Economic Entomology* 29 (1936): 808–9.
62 See J.T. Jardine *et al.*, "Report of the committee on experiment station organization and policy", *Proceedings of the Association of Land-Grant Colleges and Universities*, 42 (1928): 203–5, on p. 204.
63 University of Minnesota Archives: group ALD9E.1: fold. 32.
64 University of Minnesota Archives: group ALD9E.1: fold. 460.
65 Chapman's work started a long and continuing tradition of studying population dynamics by examining the interactions of Flour Beetles; see G. Mitman, *The State of Nature: Ecology, Community, and American Social Thought, 1900–1950* (Chicago: University of Chicago Press, 1992), pp. 127–9.
66 University of Minnesota Archives: group ALD9E.1: fold. 270 & 441.

67 See D.B. Scott & E.H. Karr, "The influence of insecticides added to soil on growth and yield of certain plants", *Journal of Economic Entomology* 35 (1942): 702–8; and J. Whorton, *Before Silent Spring: Pesticides and Public Health in pre-DDT America* (Princeton: Princeton University, 1975), pp. 212–47.

68 T.J. Headlee, "To what extent may organic insecticides be used as substitutes for arsenicals", *Journal of Economic Entomology* 27 (1934): 148–56.

69 J.S. Houser *et al.*, "Report of the committee on coordination of entomology with the war effort", *Journal of Economic Entomology* 35 (1942): 303–5. See also J.H. Perkins, "Reshaping technology in wartime: the effect of military goals on entomological research and insect-control practices", Technology and Culture 19 (1978): 169–86.

70 S.A. Rohwer, "Report of the special committee on DDT", *Journal of Economic Entomology* 38 (1945): 141.

71 See P. DeBach, "An insecticide check method for measuring the efficacy of entomophagous insects", *Journal of Economic Entomology* 39 (1946): 695–7; and A.E. Michelbacher *et al.*, "Increase in the population of *Lecanium pruinosum* on English Walnuts following the application of DDT sprays", *Journal of Economic Entomology 39* (1946): 812–3.

72 R.S. Woglum *et al.*, "The effect of field-applied insecticides on beneficial insects of citrus in California", *Journal of Economic Entomology 40* (1947): 818–20.

73 See A.E. Michelbacher, "The importance of ecology in insect control", *Journal of Economic Entomology* 38 (1945): 129–30.

74 Michelbacher, "Ecology", p. 129.

75 P.N. Annand, "Preventive entomology", *Journal of Economic Entomology* 40 (1947): 461–8, on p. 461.

76 S.A. Graham *et al.*, "Report of the committee on the relation of entomology to conservation", *Journal of Economic Entomology* 40 (1947) 150. See also G.C. Decker, "Agricultural applications of DDT, with special reference to the importance of residues", *Journal of Economic Entomology* 39 (1946): 557–62; Decker *et al.*, "Report of the committee on the relation of entomology to conservation", *Journal of Economic Entomology* 39 (1946): 695–7; and C. Cottam & E. Higgins, "DDT and its effect on fish and wildlife", *Journal of Economic Entomology* 40 (1947): 461–8.

77 D. Hounshell & J.K. Smith, *Science and Corporate Strategy: DuPont R&D, 1902–1980* (Cambridge: Cambridge University Press, 1988), pp. 451–64.

78 C.B. Williams in E.H. Smith, "Report on the XIth International Congress of Entomology", *Bulletin of the Entomological Society of America* 7 (1961): 83–92, on p. 91.

79 W.R. Henson in Smith, "Report", p. 92.

80 See C.E. Palm, "The growing responsibility of entomology to human welfare", *Annals of the Entomological Society of America* 47 (1954): 1–8;

G.C. Decker, "All entomology is economic", *Bulletin of the Entomological Society of America* 2 (1956): 2–5; B.A. Porter, "What have we learned from the Codling Moth?" *Bulletin of the Entomological Society of America* 3 (1957): 4–7; and H.M. Armitage, "Man in an insect world", *Bulletin of the Entomological Society of America* 4 (1958): 2–5.

81 See S.B. Freeborn, "The training of a professional entomologist", *Bulletin of the Entomological Society of America* 2 (1956): 23–4; and B.A. Porter, "Scarcity of entomologists", *Bulletin of the Entomological Society of America* 2 (1956): 26.

82 See C.W. Sabrosky, "Taxonomic entomology in the U.S. Department of Agriculture", *Bulletin of the Entomological Society of America* 10 (1964): 211–20.

83 R.C. Smith, "Is entomology zoology?" *Journal of Economic Entomology* 40 (1947): 608–9, on p. 609.

84 See E.G. Linsley, "Consolidation of the Entomological Society of America and the American Association of Economic Entomologists", *Annals of the Entomological Society of America* 45 (1952): 359.

85 Howard, *History*, p. 175; see also H.M. Hatch, "Entomology in search of a soul", *Annals of the Entomological Society of America* 47 (1954): 377–87.

86 L.M. Peairs, "Hash and rehash", *Journal of Economic Entomology* 39 (1946): 679–81, on p. 680.

87 R.C. Smith, "Definition of an entomologist", *Bulletin of the Entomological Society of America* 3 (1957): 4; and K.L. Knight, "Elements of entomology: The man", *Bulletin of the Entomological Society of America* 4 (1958): 42–4.

88 See C.B. Huffaker *et al.*, "The natural enemy component in natural control and the theory of biological control" in Huffaker (ed.), *Biological Control* (New York: Plenum Press, 1971), pp. 16–67; and D.P. Jones, "Agricultural entomology" in R.F. Smith *et al.* (eds), *History of Entomology* (Palo Alto: Annual Reviews, 1972), pp. 307–32.

89 J.E. Wille, "Biological control of certain cotton insects and the application of new organic insecticides in Peru", *Journal of Economic Entomology* 44 (1951): 13–8; and T. Boza Barducci, "Ecological consequences of pesticides used for the control of cotton insects in the Cañete valley, Peru" in M.T. Farvar & J.P. Milton (eds), *The Careless Technology* (New York: Natural History Press, 1972), pp. 423–38.

90 Ken Hagen (University of California), interview with Paolo Palladino, 6 July 1988.

CHAPTER 2

Entomology in Canada

CANADIAN ENTOMOLOGY, ABOUT 1850

The early history of entomology in Canada was not very different from that of entomology in the United States. Throughout the early nineteenth century, a number of naturalists had been studying the insects of what was then known as British North America.[1] By the middle of the century, some of them had become so interested in insects, and so confident of their distinct identity as entomologists, that, in 1863, they decided to band together and form the first Entomological Society of Canada.[2]

Like the Entomological Society of Pennsylvania, this Canadian society was led by a small group of metropolitan experts in insect taxonomy and systematics, busy identifying the many specimens sent to them by more untutored surveyors and settlers in the remoter parts of British North America.[3] Chief among these experts were Reverend Charles Bethune, William Saunders, a pharmacist in London, Ontario and James Fletcher, a Governmental Accountant in Ottawa. Just like their colleagues in the United States, they were tied together by a shared interest in all that the systematics of insect species could tell about Divine Providence.[4] In fact, when Canadian and American entomologists came together at the meeting of the American Association for the Advancement of Science in 1872, the highlight was a dispute between Bethune and his American opposite, Samuel Scudder, over the implications of organic evolution for this theistic vision of entomology.[5] Even as late as in the 1910s, when challenged by an aspiring economic entomologist about the utility of his butterfly collections, Bethune, then professor of entomology at Ontario Agricultural College, argued for the importance of butterflies by responding simply that "they were created because the Lord has an eye on the beautiful".[6] This was, for Bethune, a sufficient answer to the student's utilitarianism. However, it was not this link with theological questions that assured Canadian entomologists, such as Bethune, a firm institutional base for the advancement of their science, but their becoming involved with the growing movement to establish agricultural work on scientific foundations, and in particular with the development of economic entomology.[7]

During the 1850s, the Bureau of Agriculture for the provinces of British North America sought to aid farmers by awarding prizes to the authors of the best studies on insect pests. The professor of chemistry at Trinity College, Toronto, Henry Youle Hind, won the first of such competitions with his "Essay on insects and diseases injurious to the wheat crops". The agricultural importance of entomology was further confirmed by the drafting of the first plans to organize the Entomological Society of Canada during a meeting of the Ontario Agricultural Association. Moreover, the members of the newborn society sought further to strengthen these links by welcoming, in 1870, provincial legislation committing the Council of the Agricultural and Arts Association of Ontario to allocate to the Society an annual maintenance grant, in exchange for reports on economically destructive insects and the management, on behalf of the Council, of an entomological reference collection. The year after the enactment of this legislation, the devastating Colorado Potato Beetle made its first appearance in Ontario, and the Council soon called on members of the young Society for assistance. Saunders and Baynes Reed were able quickly to report that the arsenical insecticide Paris Green was highly effective against the Beetle. The reward for this was yet more legislation to increase the subvention for the Society, and to formalize its ties to the provincial government. In 1871, the Entomological Society of Canada was chartered, incorporated, and re-christened as the Entomological Society of Ontario (although nominally a provincial one, this Society was to provide the main forum for all Canadian entomologists, at least until 1950, when the Entomological Society of Canada was re-established.[8]

The importance of these ties to agriculture was also confirmed by Fletcher and Saunders' involvement in the institutional development of economic entomology in the United States during the 1870s and '80s: they were founding members of the American Association of Economic Entomologists.[9] This close association between Canadian and American entomology was further strengthened as ever greater numbers of Canadian entomologists were educated in American universities, and followed the path to professional development charted by their colleagues south of the border. In fact, Fletcher pointed out in his presidential address to the Entomological Society of Ontario, in 1889, that Canadian entomologists, like their counterparts in the United States, did not owe the high fortunes which they had achieved by this time to any advances in taxonomy and systematics, but to contributions to the agricultural economy.[10] Entomology in Canada had become, firstly and foremost, an "economic" science. However, unlike American entomologists, Canadian entomologists succeeded in maintaining a greater balance between general interest in insect life and in problems of crop protection. Even though the Chief of the Bureau of Entomology in the United States Department of Agriculture, Leland Howard wrote in his history of economic entomology that Fletcher kept "Canada ... abreast of the times, largely by the adoption and assimilation of American methods", Canadian entomologists never sought that formal separation between academic and economic entomology established in the United States with the founding in 1907 of the Entomological Society of America.[11] The roots of this difference must be traced to the institutional development of Canadian entomology, a development that was shaped profoundly by both the organization of Canadian government, and Canadians' struggle to create a national identity that would affirm their independence from both Great Britain and the United States.

THE AGRICULTURAL DEVELOPMENT OF CANADA

The history of agriculture in British North America seems to have been very different from that of American agriculture. Until the late 1840s, British mercantile policies in British North America meant that the primary function of farming was simply to produce enough food to sustain the employees of mercantile companies, such as the Hudson's Bay and Northwest Companies, which were working to export to Britain all the furs, fish, and lumber that the North American continent would produce, as well as the military garrisons necessary to protect these companies from the natives and the not very friendly Americans. Even during the 1850s and '60s, the formative years of the Dominion of Canada, the growth of agriculture in British North America continued to be subordinated to that of the economic sectors dearest to metropolitan and imperial authorities. For example, the construction of canals and railroads along the St. Lawrence and in western Ontario aimed to benefit metropolitan and British investors by facilitating the transport to eastern seaports, and to Britain thereafter, of the agricultural produce of American rather than Canadian farms.[12]

As the effort to channel American produce along Canadian trade routes began to falter, it became clear that the investments in building these routes could be made to pay only by encouraging Canadian agricultural productivity. It was agreed that more settlers should be encouraged to come to the provinces of British North America, and to take up agricultural work. There were, however, too many farmers seeking to carve their livelihood out of the limited and increasingly exhausted eastern farmlands to hope that any new immigrants might resist the enticements of American land agents and settle in Canada rather than the United States. Metropolitan investors began therefore to turn their eyes on the Northwest Territory, the then largely uncharted lands stretching from the Laurentian Shield to the western edge of what is now Alberta, and north toward the tundra.[13]

The Northwest Territory was owned, however, by British mercantile companies, chiefly by the Hudson's Bay Company. These companies were interested solely in the extraction of the abundant natural resources of the region, and very often had sought to discourage agricultural settlement, fearing that it might result in their losing control over these resources. Yet, during the 1860s and '70s, the Monroe Doctrine justifying American westward expansion was increasingly thought to apply to this Northwest Territory as well. Everyone who wielded any economic power in British North America (including the Hudson's Bay Company) realized that American encroachment upon the bases of their power could be arrested only by transforming the Territory into a land settled by colonists loyal to the British Crown. Thus, following the British North America Act of 1867, which established Canada as an autonomous Dominion of the British Empire, the new Dominion government acquired the Northwest Territory from the Hudson's Bay Company and rapidly proceeded to foster its settlement.[14]

This settlement was encouraged in 1872 by the enactment of the Dominion Lands Act, the Canadian equivalent of the Homestead Act, which had helped to settle farmers on the western lands of the United States. Railway companies also received rich commissions and land grants to build tracks from the eastern metropolis, across the newly acquired northwestern territories, to British Columbia, and thus help to improve the commercial prospects of western agriculture. Settlement was also encouraged by federal support for surveys such as that by Henry Youle Hind and George Dawson to establish and advertise the agricultural potential of the Northwest Territory. Youle Hind was recruited partly for the knowledge shown in his prize winning entomological essays, so that he might report back on any potentially destructive insects encountered during the survey. Support also went to the expansion of the agricultural improvement schemes already organized during the 1830s by the various provincial governments as palliatives against farmers' demands for a more serious effort to foster colonial agriculture. However, many immigrants attracted to Canada during this period still preferred to move on, and settle south of the border because the terms of Canadian settlement legislation were far less liberal than those in the United States. By the 1880s, this policy was clearly counter-productive; as long as there were no settlers, the railway companies failed to make any profit from either the railways or the land they had received in payment for building the railways, and still there was no answer to the American threat.[15] The Dominion government sought therefore to make the terms of homesteading legislation more generous, mostly by opening more of the attractive rail-side Crown Lands for sale. It also sought to encourage the establishment of vast irrigation schemes wherever the lands of the Northwest Territory and Manitoba might seem uninviting to any potential settlers.[16]

These initiatives, and the many other ones to facilitate settlement, began well in advance of the arrival of any substantial number of settlers; it seems, therefore, that the settlement of western Canada was a much more directed affair than in the United States. The historian Vernon Fowke has, in fact, argued that the policies for the agricultural development of British North America, and later for the development of the Dominion of Canada, were tailored to meet the needs of metropolitan and transatlantic commerce to a far greater extent than in the United States.[17] While this sharp differentiation should be accepted with some caution, especially in view of more recent, convergent scholarship on the settlement of the American and Canadian western lands, some of the elements involved seem quite useful to understand the different development of Canadian entomology.[18]

THE DOMINION EXPERIMENTAL FARMS

Among their many different goals, the surveys of the Northwest Territory aimed to dispel the notions promoted among immigrants, firstly by the Hudson's Bay Company, and then by American land agents and land speculators, that the Territory would never allow profitable farming.[19] Enormously destructive plagues of the Western Grasshopper in the prairie regions of Canada during the 1870s provided particularly effective material for this campaign to deter their settlement. The Dominion government sought to dispel discouraging news about the grasshoper by aiding Canadian entomologists' collaboration with Charles Valentine Riley's United States Entomological Commission, which was undertaking at this time an extensive study of the problem south of the border.[20] George Dawson and Henry Youle Hind were soon able to suggest in their reports on the situation that farmers could halt the crawling hordes of locusts by flooding trenches around their fields — an unusually clear example of the amount of work often required by cultural methods of crop protection.

This support for entomological work on the Grasshopper was eventually translated by the Dominion Department of Agriculture into the appointment, in 1883, of a Dominion Entomologist, James Fletcher, to advise Canadian farmers on any other problems they might have with insects attacking their crops.[21] During the next couple of years, Fletcher established, through his extensive and assiduous correspondence with these farmers, a network of over four-hundred entomological observers.[22] These observers were then moved by Fletcher and his colleague in the Entomological Society of Ontario, William Saunders, to build pressure among the various associations promoting agricultural improvement across the Northwest Territory for more systematic governmental support of such improvement. In 1885, Saunders was asked by the Dominion Parliament to prepare a report comparing the organization of the agricultural colleges, experimental farms, and experimental stations in the United States and Canada, a report which might provide the basis for the establishment of a Canadian national organization for agricultural improvement.[23]

Saunders' report to the Minister of Agriculture, John Carling, fully illustrates how impressive was the scale of American efforts to establish agricultural science and education. However, Saunders was not interested in promoting this example in all its details, for he argued that what Canada needed was a system to support agricultural science alone. He viewed the agricultural colleges as not very effective as they drew few students, and tended to impart a general education rather than any immediately practical instruction in farming.[24] It is more probable, however, that the recommendation against any conjunction of colleges and experimental farms or stations was based on the realization that it would have been politically impossible. The two activities had to be kept separate because the British North America Act stipulated that the federal government was responsible for the use of national economic resources, and that matters of social policy, such as education, should to be left to provincial authorities. Since the research organization proposed by Saunders was intended to improve the uses of what was by the late nineteenth century a very important Canadian natural resource, its farmland, support for agricultural research was more easily construed as a federal responsibility than was possible with agricultural education. Saunders also argued in his report that this organization should be centrally controlled because a decentralized one such as that in the United States would only disperse very inefficiently all too precious and limited financial resources.[25]

Notwithstanding the carefully embodied political precautions, the legislation to establish and fund a national center for agricultural research, which the Minister of Agriculture presented to the Parliament on the basis of Saunders' report, was strongly opposed by the editor of the influential *Farmers' Advocate and Home Magazine*, William Weld.[26] He argued that Canadian farmers did not need any experimental farms because they could easily rely on the results of work conducted in the United States. In fact, the "dry farming" techniques recommended by Saunders to develop the Northwest Territory could easily have been borrowed directly from the United States.[27] If a Dominion Experiment Farm had to be established, Weld and others argued, it was better that it should be organized in a manner similar to the agricultural experiment stations in the United States. That is to say, there should be a number of farms controlled by the government of the provinces in which they were located. Just as in the United States, this opposition was largely motivated by concern that the establishment of the Dominion Experiment Farm was just another example of federal interference in affairs that were better left to private enterprise, if not an example of corrupt metropolitan political patronage. However, Weld's esteem for Saunders' high scientific reputation seems to have appeased him. Weld admitted eventually that Saunders was,

> not an office seeker. He is an eminent chemist who can make an honest living, and has devoted his spare moments to various sciences out of pure love for them. He is not a practical agriculturist, and although the science of agriculture is not his forte, yet he is an eminent horticulturist and entomologist, and we believe that he could easily adapt himself to the new situation, should he be appointed director of the experiment stations ...[28]

In these words we can read many of the more practically oriented views that shaped the evolution of entomology in the United States, which might still lead us to expect that it would develop very similarly in the two countries.

A compromise between Weld and Saunders' perspectives was reached in 1887, when the Dominion Parliament agreed to establish one central Dominion Experiment Farm for Ontario and Quebec, and four more branch farms across the other provinces, all under Saunders' direction. During the next two decades, the latter proceeded further to defuse the tension between federal authority and provincial interests by establishing a number of other, subsidiary experimental farms.[29] Importantly, however, these new centers were often established, like the railways and irrigation canals built supposedly to serve the Canadian farming community, well in advance of the settlers' arrival. They were created chiefly to meet the needs of metropolitan political and economic interests, rather than any local communities. Furthermore, when the establishment of the individual experiment farms was initiated by local commodity groups or provincial authorities, they then came under the control of a distant, federal organization in which leading scientists enjoyed considerable power to shape the development of the farms according to their own preferences (and those of the director of the Central Experiment Farm in Ottawa) rather than those of their local sponsors.[30]

The position of the scientists in the experimental farms was, in sum, more akin to that of scientists in the United States Department of Agriculture than to that of scientists in the state agricultural experiment stations. Perhaps they even enjoyed more power than the former group of American scientists. Charles Saunders, William's son and Dominion Cerealist, complained in 1919, before the Cronyn Committee, which was then reviewing the organization of government funding for scientific research, that scientists in the Dominion Department of Agriculture had few opportunities for research because they were unduly constrained in their work by their formal responsibility to the farming community. However, ever since the arrival in Canada of the San José Scale in the late 1890s, entomologists in the Department enjoyed great power to quarantine infested areas, and to order farmers to spray or destroy crops and orchards as needed to control the spread of any noxious insect species, even if the farmers were opposed to any of these measures.[31] In the United States, entomologists had to fight quite hard to establish any such quarantine measures, and, even as late as in the 1920s, these American measures were viewed by Canadian entomologists as very weak, if not "worthless".[32] This greater authority in relation to the farming community was very important for the development of Canadian entomology during the early twentieth century into a quite different science from entomology in the United States.

ENTOMOLOGY AND THE ORGANIZATION OF AGRICULTURAL SCIENCE

For all his scientific merits, William Saunders was chosen to conduct the first exploratory investigation of the American organization of agricultural science and education chiefly because he was politically well connected with the metropolitan political elite. He and the Minister of Agriculture, Carling, were fellow members of the Entomological Society of Ontario, neighbors in London, Ontario, and very good friends.[33] The turn of the century, however, was marked by a revolt of the urban, professional middle classes against a moral degeneration of the Canadian people, which led, among other problems, to the patronage in government and civil administration that made Saunders' appointment possible.[34] These concerns about degeneration were certainly heightened by the rapid influx of alien immigrants from eastern and southern Europe to settle the mainly agricultural western provinces, who often failed to integrate smoothly into a different, more commercially oriented organization of agricultural production, and played an important role in rising rural discontent.[35] Canada, reformers from these professional middle classes argued, had to be rebuilt through a more efficient management of its human (and natural) resources into a modern nation, and science was an important component of this effort. Thus, the entomologist and eugenist William Lochhead, at MacDonald College, the college of agriculture attached to McGill University, believed that an education in entomology was crucially important for the education of more civil farmers. He wrote in the *Annual Report of the Entomological Society of Ontario* that,

> entomology must be presented in such a way that it will help … to protect crops. At the same time we should not forget that entomology is a cultural subject, and should be a means of training the young men to observe carefully, to experiment, and draw conclusions. It should also give them an added interest in the great world of nature about them, and enable them to get a glimpse into the wonderful web of life, with all its marvelous inter-linkages. This phase of the subject is, in my opinion, one of the most important from the standpoint of citizenship. From a study of the insect world it is an easy matter to direct their attention to inter-linkages in the social world.[36]

This call for education to build a crop of efficient, "progressive" farmers, was supported by the enactment in 1913 of the Agricultural Instruction Act, which allocated $10 million for the establishment of agricultural colleges across the nation. The Dominion Experiment Farms were an equally important element in this plan. They rapidly increased in size and number.

The staff to fill the newly created positions in agricultural education and research was, however, to be selected in a manner quite different from that of previous decades. The Civil Service Act of 1908 stipulated that, from now on, anyone who sought employment in the federal government, including scientists, had to be selected on the basis of their professional accomplishments rather than their political connections.[37] This meant that the earlier generation of self-educated experts such as Saunders and James Fletcher was to be replaced by professionally trained scientists.

Canadian universities, however, were not yet preparing sufficient numbers of scientists to fill the new positions. The vast majority of scientists recruited into the expanding Dominion Experiment Farms were, therefore, trained in either Britain or the United States, where they were often converted to the tradition of scientific research increasingly common in these two countries from the 1890s onwards. Many of them believed that what Mark Slingerland in the United States had called the "routine" work of simply identifying insects attacking agricultural crops, and then devising or improving means to control them, was an inefficient use of their expertise. They should instead be allowed to focus their attention on the most general aspects of insect biology necessary to establish truly effective control measures. One advocate of this view was the graduate of the University of London, and Dominion Chemist, Frank Shutt, who, in 1919, told the Cronyn Committee that,

> there must be an opportunity for continuous and uninterrupted patient and free work ... The investigator ought to be as free as possible from routine work, and from the rules and regulations found necessary in the prosecution of routine work.[38]

Shutt may have been reflecting upon his own work on, among other things, the testing of insecticides and the arsenical residues they left on Canadian apples, which had led to the banning of this produce from the British market.[39] Shutt's line of argument was further pursued, on the same occasion, by the Dominion Cerealist, Charles Saunders, holder of a doctoral degree in chemistry from the even more prestigious Johns Hopkins University. He argued that agricultural scientists should be well isolated from any political or bureaucratic pressure, and that, therefore, the Dominion Experiment Farms should be removed from control by the Department of Agriculture.[40]

While Shutt and Saunders' views were largely rejected by the Cronyn Committee, federal entomologists were well on their way to the independence they advocated. They had the funds: out of the $10 million allocated by the Agricultural Instruction Act to agricultural education, the very considerable sum of $620,000 went to entomological investigations alone.[41] They were also institutionally independent: in 1914 the Department of Agriculture separated the entomologists from all the other agricultural scientists it employed by establishing an Entomology Division quite apart from the Experiment Farms.[42] Moreover, as we have already seen, they had much power to overrule farmers' stated interests and impose measures such as quarantines against foreign insect pests.

Much of this special position for entomology was the result of skillful work of the second Dominion Entomologist, Gordon Hewitt. Not only was he particularly active in the development of the Agricultural Instruction Act and the incorporation into it of a recommendation to foster "instruction in horticulture and entomology", but he also used the exceptional powers to inspect and quarantine, vested in him by the Destructive Insect and Pest Act of 1910, rapidly to expand the newly established Entomology Division. Hewitt argued that these latter responsibilities could not be met in any adequate manner with the limited personnel of the Division, and thus the number of entomologists it employed had to be increased quite dramatically.[43] Hewitt's entomologists then went to work on a number of problems beside those caused by the San José Scale and other exotic pests, and, by 1915, they were at work in nine new entomological field stations scattered across Canada.

Hewitt also worked to change the orientation of federal entomologists' work. For example, from his very arrival in Canada he objected to his colleagues' preoccupation with the development of insecticidal methods, and recommended instead work on biological control. He wrote in the *Annual Report of the Entomological Society of Ontario* that,

> our knowledge of the natural means of control, the parasitic insects, and also the parasitic fungi, which under natural conditions frequently obtain control, is still very meager. One reason for this is, I think, that economic entomologists have been too much occupied in devising sprays and other mechanical means of attacking injurious insects, and the study of parasitic means of control, the natural means of control, has been unduly … neglected.[44]

As befit a graduate of the leading department of zoology at the University of Manchester, Hewitt's was not simply a call to increase the introductions of predators and parasites, but a call to improve the effectiveness of biological control by closer study of the bionomics and ecology of insect species in natural as well as artificial environments. The Entomology Division was to become a center for research on the general biology of insects, not just a home for specialists in the development of better chemical and biological "mousetraps"; Hewitt had little time for specialists in economic entomology, and instead looked to recruit people who knew only so much entomology as was "consistent with other classes of the arthropods in the basic course of zoology".[45]

Hewitt's supercilious attitude toward strawberry growers in the Okanagan Valley of British Columbia captures well the transformation he had brought about by the end of his decade as Dominion Entomologist. In 1918, he was called upon by these growers to help them protect their crop against the Strawberry Root Weevil. They soon grew frustrated with his and other entomologists' focus on the life history of the Weevil, and with their inability to do anything more concrete to aid them; when one such grower confronted Hewitt and his colleagues, questioning their claims about the utility of these investigations, Hewitt responded by threatening,

> we are not here to listen to remarks of that description and we will withdraw immediately from this investigation unless you accept what we tell you as the truth.[46]

This uncompromising response, which presumed entomologists' institutional independence from the growers, would hardly have been imaginable in the United States. Perhaps things like this also happened there as well, but they certainly were not recorded in any of the main American journals of economic entomology.

ENTOMOLOGICAL WORK IN THE PROVINCES

Even after the enactment of the Agricultural Instruction Act in 1913, provincial governments were so much poorer than the federal government that they could not easily afford to encourage the development of agricultural colleges on a par with the land grant colleges in the United States.[47] As William Lochhead pointed out in a survey of graduate education in entomology, Canadian agricultural colleges were organized quite differently from American ones, as they divided their courses in two streams, a two year one for a diploma and a four year one for the degree in agriculture (a plan which closely resembled that followed in Great Britain). Most students followed the former stream and then returned to the farm. Only a few went further than the introduction to insect ecology (and all that it taught about civility) recommended by Lochhead, to follow the more technically advanced degree course, in the hope of eventually gaining thereby teaching posts in schools, colleges, and universities.[48] Moreover, any support for research in these institutions was quite out of the question. Thus, many Canadian students followed more advanced, graduate courses either in Britain or the United States.[49] Most of them chose the United States, and thus reinforced a rapidly growing tie between Canadian and American entomology.

Following a well established tradition that the Canadian scientific elite should be drawn from scientists allied with the British rather than the American scientific community, however, many senior entomologists in the Entomology Division of the Dominion Department of Agriculture, such as Gordon Hewitt, were recruited from British universities.[50] This preferential policy sometimes frustrated American entomologists such as Leland Howard, in the United States Department of Agriculture. He was consulted during the search for a successor to first Dominion Entomologist, James Fletcher. Howard's preference was for the latter's assistant, Arthur Gibson, but he believed that "there was a feeling … that the Dominion should rely on the old country as much as possible in such matters".[51] While Gibson eventually succeeded Hewitt, who died of influenza during the terrible pandemic of 1920, the far larger pool of entomologists trained in the United States began their Canadian careers in provincial organizations.[52]

Some attempts were made in 1914 to coordinate the activities of federal and provincial organizations supporting agricultural research or otherwise technical work in agriculture by organizing occasional coordinating conferences, but they do not appear to have been very effective; by 1935, federal and provincial agricultural scientists constituted separate communities with very different responsibilities and interests.[53] For example, since the 1920s, the main brief of federal entomologists had become research on the systematics and life histories of economically important insect species. Provincial governments focused their support, instead, on the more immediately important organization and implementation of crop protection programs. Thus, the Provincial Entomologist for Ontario, Lawson Cesar, was mostly busy training entomological inspectors, teaching farmers about spraying methods, and preparing spray calendars; his more lasting contribution to Canadian entomology was his *History of Spraying in Ontario*.[54] Federal entomologists in some of the experimental farms across the provinces, such as those at the Annapolis Royal Laboratory, Nova Scotia, also followed this path. In the 1920s and '30s, local, commercial apple growers were particularly vocal about their dissatisfaction with not receiving sufficiently "definite and timely information about spraying" and had proceeded to undermine these entomologists' institutional position by creating, with some assistance from the Provincial Department of Agriculture, their own "spray circles". These conflicts were reported in Arthur Gibson's unpublished history of economic entomology in Canada, written in 1951, which, very interestingly, is generally defensive about the contribution of federal entomologists like Gibson himself to the work of their counterparts in the provinces.[55]

More significantly, since the use of chemical insecticides had been presented to farmers as the "progressive" way to tackle insect problems ever since the eradication campaigns against the San José Scale, and because companies selling them soon provided a more extensive advisory network than the provincial authorities could afford, farmers grew increasingly committed to chemical control. By the mid-1920s, chemical insecticides were as popular in Canada as they were in the United States.[56] It is not surprising, therefore, that the work of entomologists in the provinces tended increasingly to focus on the improvement of chemical control programs. Peter Venables, a federal entomologist in British Columbia, suggested rather sarcastically that, in the horticulturally rich Okanagan Valley, these entomologists usually leapt around spray tanks, brandishing spray guns.[57] By the end of the 1930s, Canadian apple growers, like their American counterparts, were confronting serious problems as a consequence of this commitment: Britain banned their fruit just like that from the United States for being contaminated with unsafe levels of arsenical insecticides.[58]

The entomologists advising farmers on the use of chemical insecticides did not, however, enjoy the benefits that accrued to their American colleagues from their commitment to chemical control. The director of the Science Service of the Department of Agriculture, Kenneth Neatby, noted in 1951, that historically most insecticides used in Canada had originated (and were still produced) in the United States.[59] This assessment is amply confirmed by the most casual perusal of provincial publications such as the *Annual Reports of the British Columbia Entomological Society*, which are full of advertisements for insecticides manufactured by Chemagro, Dow, Niagara, Röhm & Haas, Sherwin Williams, and Stauffer. Unfortunately for Canadian entomologists, these companies, just like most other purveyors of industrially manufactured goods sold in Canada, relied on research in American rather than Canadian academic or industrial laboratories to develop more economic and effective products. Furthermore, since governmental funding for research went to governmental laboratories rather than academic departments, there would have been few opportunities for either American or Canadian chemical companies to support research there even if they had been interested in doing so. This lack of industrial support was a serious source of concern for the Canadian government, because it resulted in a substantial emigration of scientific manpower to the United States, and increasing scientific, technological, and economic dependence of Canada on its southern neighbor.[60]

SEPARATING RESEARCH AND OPERATIONAL WORK

In Canada, the 1930s were marked by increasing centralization of governmental power, and governmental intervention in all spheres of economic activity, partly in response to the Great Depression. The Canadian agricultural research system was expanded even further to confront the challenge of lowering farmers' production costs so that they might better succeed in an increasingly protectionist world market. This growth was also accompanied by a fundamental reorientation of federal agricultural research.

Like their counterparts in Britain, agricultural scientists in the Dominion Experiment Farms sought to escape oversight by the Department of Agriculture, and to become the main center for Canadian research in the life sciences. The effort was largely unsuccessful due to the combined resistance of the Minister of Agriculture, James Garddiner, and representatives of the most important Canadian scientific institution, the National Research Council. While the former was always suspicious of anything but the most practically oriented scientific work, and also believed that Canada could very well rely on American work, the latter feared that any institutional rearrangement might undermine its own great institutional independence.[61] However, the generation of Canadian scientists who had been trained in the United States during the first years of this century, began increasingly to enter more important administrative positions. In so doing, they imported into the upper levels of the Canadian agricultural research system the ideal of specialized agricultural research which blossomed during the interwar period in the more prestigious American land grant universities such as Cornell University and the University of Minnesota.

The Entomology Division was deeply affected by these developments as federal entomologists' views that they could be more effective only if given greater opportunity to pursue more sustained, ecological lines of work began to find greater institutional support. In 1932, James Swaine, a former student of John Henry Comstock at Cornell University, who had been brought into the Entomology Division by Gordon Hewitt some twenty years earlier to work on the ecology and biological control of the Larch Sawfly, became Director of Research of the Canadian Department of Agriculture.[62] From this advantageous position, he sought to extend Hewitt's vision of the proper relationship between research and operational work to all areas of agricultural science. He proposed a formal and general separation of agricultural research and more operational work by dividing the activities of the Department of Agriculture into four administratively separate organizations, one each for production, marketing, experimental farms, and science.[63]

In 1937, Swaine's plan was accepted by the Department of Agriculture, and it brought to the newly established Science Service what a former scientist in, and official historian of, the Department has called an "Ivory Tower" approach.[64] Thus, for example, the poisoning of bees in Nova Scotian apple orchards by growers' intensive use of arsenical insecticides, which caused the apple blossoms on many trees to wilt unpollinated and thus reduce the trees' productivity, had to be studied "from a fundamental standpoint".[65] This meant that when, in 1939, Allison Pickett moved from the office of Provincial Entomologist for Nova Scotia, which under his tenure had established and supervised very successful, compulsory spraying programs against the Apple Maggot, to the post of director in the entomological laboratory at Annapolis Royal, he immediately set out to de-emphasize the more empirical work on chemical insecticides, and then organized a research program to study the ecology of the four major pests in Nova Scotian apple orchards, the Oystershell Scale, the Red Mite, the Eye Spotted Bud Moth, and the notorious Codling Moth.[66]

EXPANDING THE DEFINITION OF RESEARCH

This institutional and intellectual reorientation of the Canadian agricultural research system was reinforced in the years after the second world war. Moreover, thanks to an economic boom, and Canadians' changing perceptions of their place in the world as more than just an appendage of the British Empire, the pace and direction of change was far more rapid and radical than ever before. For example, the National Research Council had been established, during the first world war, to assist the industrial development of Canada by supporting scientific work related to economic goals. During the second world war, the brief and powers of what had already become the principal Canadian center for scientific research were radically transformed. Leading Canadian scientists, such as Edward Steacie, were handed great latitude to shape their involvement of the national war effort.[67] They were frustrated, however, by the subordination of their programs to British and American planning. After the war, aided by a political system that was extremely sensitive to the vagaries of personal relationships between political figures such as Clarence ("C.D.") Howe, the influential architect of Canadian postwar reconstruction, and administrative bureaucrats, such as Howe's erstwhile student and then president of the National Research Council from 1939 to 1952, Chalmers Mackenzie, these scientists took the initiative to expand the horizons of Canadian support for scientific research. It also helped that both Howe and Mackenzie had been trained as scientists before entering public administration; for them too, providing greater support for Canadian scientific research was one of the most important ways to bolster Canadian national identity and independence. The National Research Council actively began to encourage a greater emphasis on the most general scientific aspects of industrial, or otherwise practical activities, and its laboratories were increasingly organized along the lines of academic ones.[68]

Significantly for the history of entomology in Canada, Howe took keen interest in the establishment in 1944 of the Forest Insects Control Board and the Forest Insects Laboratory at Sault Ste Marie, Ontario.[69] The lumber industry was very important in Canada. In 1927, Canada produced two thirds of the world's wood pulp, its main export commodity, and, more significantly for the war effort, the famous Mosquito fighter airplane was made of Canadian plywood.[70] The Spruce Budworm and Larch Sawfly attacking Canadian forests together became, therefore, the Canadian equivalent of the Cotton Boll Weevil. Unlike cotton plantations, however, the use of insecticides over these immense forests was absolutely prohibitive until the advent of DDT. All that could be done was to ensure alertness to local infestations, and treat them before they became major outbreaks. Effective forecasting was, therefore, extremely important, and understanding the population dynamics of these insects was a major preoccupation for Canadian forest entomologists. The laboratory in Sault Ste Marie soon became a major center for such ecological research, as well as one of the leading in the world for research on insect pathogens and their uses in pest control.[71]

More significantly still for Canadian entomology, research in the newly renamed Canadian Department of Agriculture was further transformed during the post-war boom. There was widespread concern that Canadian agricultural research was not having any impact on the national agricultural economy, which motivated many debates over the relationship between "basic" and "applied" science. Leading scientists and administrators, such as the two graduates of the University of Minnesota, former associates in the very prominent Cereal Rust Research Laboratory in Winnipeg, Manitoba, and then directors of the Science and the Experimental Farms Services, Kenneth Neatby and Cyril Goulden, believed that the reason for this failure was that inadequate attention was being paid to the relationship between scientific work and the development of associated technologies. These two activities had been formally separated in 1937, the former being confined in the central, metropolitan laboratories, while the latter was made into an activity for the experiment stations across the provinces. However, it was commonly observed that provincial centers such as the Cereal Rust Research Laboratory increasingly supported work that seemed important to sustain their formal responsibilities to farmers in the Wheat Belt, but was usually conceived as an exclusive prerogative of the central laboratories. This, Neatby, Goulden, and other similarly minded administrators, such as another graduate of the University of Minnesota, the entomologist Robert Glen, argued, was an untenable situation because it led to much inefficient duplication of work. They felt that, if the Canadian agricultural research system really was to accomplish its economic mission, it had to be reorganized once again.

These reformers argued that some modified version of the American organization of agricultural research should provide the model for restructuring and modernizing agricultural research. More specifically, Neatby thought that the administrative divisions between the American federal bureaus and the diverse state agricultural experiment stations were bound to lead to wasteful duplication, which could, however, be avoided through much stronger central control over the direction of, and funding for agricultural research. The Bureau of Animal Industry of the United States Department of Agriculture, a central institution, integrating a number of distinct scientific disciplines around a particular economic mission, seemed just the kind of organization needed. This preference for centralized, integrated control was reinforced by contacts with Sir William Slater, secretary of the British Agricultural Research Council, also a highly centralized organization notable for its independence from agricultural producers, even as represented by the Ministry of Agriculture. Significantly, this independence of British agricultural scientists resulted in work more inspired by academic goals than by the needs of the agricultural producers it was intended to serve. After a series of long negotiations, Slater's recommendation to merge Neatby's Science Service and Goulden's Experimental Farms Service was taken up, and a new Research Branch was established in 1958.[72]

Entomology followed the pattern set by the Department of Agriculture. By the late 1950s, there were three times as many entomologists in the Department as there had been just after the war. These entomologists constituted over two thirds of the five-hundred professional entomologists employed in Canada, and according to Glen, at the time head of the Division of Entomology, they "conducted at least 85 percent of the entomological investigations in Canada".[73] About two thirds of the remaining entomologists found work in provincial organizations, and the rest, about ten percent of all Canadian entomologists, were employed by university departments. Entomology was, to use Glen's own words, an "applied science".[74] Significantly, he did not turn to the more usual adjective adopted in the United States, "economic". In fact, the orientation of this science seems to have changed quite drastically during the 1940s and '50s, and in a very different direction from that followed by American entomologists.

Judging from the pages of the chief American entomological journals, namely *the Journal of Economic Entomology* and the *Annals of the Entomological Society of America*, at least half of all American entomologists wrote about chemical insecticides and their uses.[75] The situation in Canada seems to have been quite different. The number of papers in the *Canadian Entomologist* on insect physiology and toxicology grew dramatically, but so did those on insect ecology. The only subjects to have suffered during this period were systematics and morphology.[76] What is even more striking about the Canadian entomologists writing these papers, especially those in the government research institutes, is that they were increasingly identifying their work with subjects other than entomology, and that these had little to do with chemistry. Instead, entomology in the Canadian Department of Agriculture, just like other organismic sciences, such as zoology, ornithology, or mammology in American universities, was disappearing as these sciences were merged and restructured into diverse sub-fields of biology. Thus, William Beckel, professor of entomology at the University of Toronto wrote in 1964 that,

> I have talked with the scientists in the research branch of the Department of Agriculture. They have problems created by insect pests that must be solved. But they ask me: "Do you have a good Biochemist, or a good Cytologist, or a good Geneticist". The rarely say anymore, as they used to: "Who are you training in Entomology?[77]

While Beckel did not refer specifically to ecology, it too loomed large among the sub-fields of biology fostered by the Research Branch. In 1951, for example, Neatby interpreted the cautions by the British physiologist and entomologist

Vincent Wigglesworth about the risks involved in using DDT, and about the need for more biological investigations, to mean that the Science Service should encourage "wider ecological study" of entomological problems.[78] This importance of ecology (and natural history) was captured by Glen, as he celebrated one hundred years of Canadian entomology by concluding that,

> Canadians might well take satisfaction from the progress attained ... But there is no cause for complacency. Most of Canada still remains to be surveyed intensively for its insects; merely a beginning has been made on the perplexing problem of the mechanisms that affect the abundance of species; the costs of control and the toxic residues on produce and in soils require continued attention; and many new entomological fronts will arise from prospective developments as the settlement of the northland, widespread industrial growth, enlargement of universities, and shifting emphases in agriculture and forestry. Fortunately, entomology in Canada is in a relatively strong position to meet the challenge of its second century.[79]

While these words suggests that the federal entomological laboratories were becoming centers for research in basic problems of biology to a far greater extent than in the United States, they also give an interesting perspective on what most worried Canadians in the late 1950s.[80]

REORIENTATION

The pace and direction of this expansion began to change as the postwar economic growth began to slow in the late 1950s. Canadians became increasingly concerned about their future. A report in 1956 by the Royal Commission on Canada's Economic Prospects suggests that they worried about the persistent failure of Canada to develop independently of the United States, both economically and culturally. Among the many questions raised was the economic effectiveness of federal investments in research.[81] These latter questions were particularly important in an influential report prepared in 1963 by the Royal Commission on Government Organization, the Glassco Report. It strongly criticized the National Research Council for being far too preoccupied with scientific research, and far too little with the development of the scientific manpower needed to foster Canadian industry.[82] These problems were also evident, albeit on a lesser scale, in the organization of the Research Branch of the Department of Agriculture.

Notwithstanding the justifications during the 1950s for the merger of the Science and Experimental Farms Services into the Research Branch in terms of improving their economic impact, in actual fact, most attention went to the problem of fostering multi-disciplinary scientific work. The separation between agricultural scientists in the federal institutes and their agricultural constituencies was actually increased. Earlier, regional research centers had helped to maintain some contact with the latter, but that contact was now much attenuated.

Moreover, the Marketing Service was not integrated in the Research Branch, but continued instead as a separate organization, thus foreclosing any link with the world of agricultural production that might have been possible through agricultural economists.[83] Tandy Johnson, the Supervisor of Agricultural Representatives for Northern Saskatchewan, complained in *Agricultural Institute Review*, a forum established in 1945 to bring together governmental organizations involved in agriculture and the agricultural industry, that research organizations should make stronger efforts to consult with their "customers" through the extension service. As he put it, "the 'manufacturing' of a product is largely the responsibility of the research worker and those who formulate policy ... improvements could be made by the 'manufacturers' in obtaining information from the 'salesman' as to how the products are being accepted".[84]

One consequence of this difficult relationship between research, development, and extension, was that Canadian farmers were just as committed to the use of DDT as any of their counterparts in the United States, even though entomologists in the central federal laboratories knew about, and warned against the dangers involved in this commitment. Their findings were simply not finding their way to the farms, orchards, and forests.

There were continuous discussions about the educational curriculum for entomologists, and the relative importance of a broad education in general biology, if not ecology, as was desired by Reginald Balch in the Forest Insect Division, versus a more specialized one in problems of economic entomology. These discussions suggest that there was a troubling disconnection between research and agricultural practice, and that it was the result of Canadian universities' failure to attend to the needs of extension entomologists. These entomologists, the vast majority of Canadian entomologists, turned increasingly to American universities for their education, and came thereby to focus almost exclusively on those aspects of entomology directly concerned with crop protection and chemical control.[85] A deep split was developing between entomologists involved in research and those involved in the more practical work of protecting crops, orchards, and forests. Thus, Peter Venables complained in 1952 that "the control of insects now seems to be in the hands of the chemical engineer, and universal destruction is the order of the day".[86] He was not exaggerating. Rachel Carson, in *Silent Spring*, described in detailed terms the massive destruction of salmon on the Miramichi River caused by government entomologists repeated aerial spraying of DDT over the forests of New Brunswick to control outbreaks of the Spruce Budworm, a destruction whose impact on public perceptions of the problem was as comparable to that of the bald eagle in the United States.[87]

ENTOMOLOGY IN THE UNIVERSITIES

Although the Royal Commission on Government Organization commented quite favourably on the work of the Research Branch, it suggested nonetheless that the reforms of the 1950s had not gone far enough.[88] More pressure was placed on the Research Branch to improve its economic effectiveness by translating its commitment to multi-disciplinary, mission-oriented work into mergers and restructuring of its various laboratories. Some scientists actually sought to resist these pressures. Other, more pragmatic ones, such as the director of the Entomology Research Institute for Biological Control, Bryan Beirne, hoped to maintain a balance between interest in more academic subjects such as Beirne's in insect systematics and bionomics, and ambitions to develop more advanced strategies of pest control such as integrated control, by moving the federal laboratories closer to universities and thus facilitate contact between scientists in the two institutions. This ultimate integration of federal laboratories and academic research departments did not take place, and there followed a massive migration of agricultural scientists from government laboratories into university departments, often in the United States.[89] The Research Branch was thus drawn into a far more radical reorganization of scientific research, as the failure of scientific research sponsored by the federal government to fulfill its economic ambitions led administrators responsible for scientific research to question whether support for research in the universities might not be a more effective policy. Canadian universities were, however, ill-equipped for this change.

During the 1920s and '30s, the increasing political and economic strength of the provinces had led provincial governments to be more responsive toward the demands of the colleges and universities for which they were responsible. Thanks to increased funding, they grew more rapidly and became more assertive about the need for more research.[90] Agricultural colleges, for example, sought to take over the Experimental Farms Service, and establish thereby a relationship similar to that between the land grant universities and the state agricultural experiment stations in the United States.[91] This, however, was resisted by both the Department of Agriculture and academics such as the entomologist John Detweiler, at the University of Western Ontario, who worried about the soundness of the American model, with all its encouragement of narrowly specialized scientific work.[92] This concern persisted into the post-war years, as the president of the Entomological Society of Quebec, George Moore of McGill University, complained that economic entomology was not a "discipline", but "that aspect of insect life which affects human affairs", and thus an education in economic entomology made little intellectual sense.[93] More significantly, the response of the Department of Agriculture meant that little agricultural research was possible in the agricultural colleges. If any graduates aspired to training at the graduate level, they might be able to obtain it in the more prestigious metropolitan universities such as the University of Toronto and McGill University, but even there facilities were poor and the training would not be strongly oriented toward problems of economic entomology. If they wanted advanced training in economic entomology or any other similarly practical subject, there was no question that, much to the frustration of academic authorities, they had to migrate south of the border, probably to Cornell University or the University of Minnesota.[94]

After the second world war, however, the universities enjoyed a period of renewed and unparalleled growth as the federal government provided subventions for the education of returning veterans. Such support continued into the 1950s as the federal government began to believe that the universities, like the federal laboratories, should become the beacons of a new Canadian cultural identity.[95] These universities, however, still lacked the research facilities needed to attract and retain both the more promising researchers necessary to establish graduate courses, and the students needed to fill the seats in these courses, because funds specifically earmarked for research continued to go almost exclusively to scientists in the laboratories of the federal government. Thus, in 1958, one third of the entomologists in the Canadian Department of Agriculture had obtained their masters degree in the United States, and the figure for doctoral degrees was one half the total.[96] Who should be doing basic, disciplinary research became a divisive issue. Moreover, entomologists in the universities knew that the vast majority of graduates would have to find employment in government laboratories, where what was wanted was specialists in the growing, new experimental areas of biology, not those schooled in the older traditions of morphological and natural historical studies.[97] However, many entomologists in both the older, metropolitan universities and in the newer, western ones, such as William Beckel at the University of Toronto and Jacob Rempel at the University of Saskatchewan, were not prepared to transform their departments into centers for specialization and professional training along the lines of the "Cornell System".[98] Ironically, both Beckel and Rempel were graduates of Cornell University, and, more significantly, Beckel had worked as a physiological ecologist in the Entomology Laboratory of the Department of Agriculture at Chatham, Ontario. Newer universities, such as Simon Fraser University, were more receptive. In 1967, for example, Bryan Beirne moved eight of his research staff at the Entomology Research Institute for Biological Control to this university, to establish the first Canadian graduate course in pest management. In sum, Canadian entomologists found themselves having to come to terms with issues which American entomologists had confronted and settled much earlier in the century. The transfer of the center of entomological research from the federal government to the universities was not going to be easy, and, even by the late 1970s, the balance between employment in universities and in governmental laboratories, federal and provincial, had not changed much.[99]

ENTOMOLOGY AND ECONOMIC ENTOMOLOGY IN CANADA

The historian Conner Sorensen has observed that during the formative years of Canadian entomology there was far less antagonism between academic and economic pursuits than in the United States.[100] He argues that this difference was the result of the later development of the Canadian entomological community;

as this community began to develop in the late nineteenth and early twentieth centuries, economic entomology was already well established in the United States and those Canadian entomologists interested in economic problems alone did not need to break ranks with academic entomologists because they could find their forum in the American Association of Economic Entomologists. As we have seen in this chapter, the situation described by Sorensen seems to have remained in many ways unchanged since the late the nineteenth century.

The insect toxicologist and director of the prestigious state agricultural experiment station in Storrs, Connecticut, Neely Turner, commented in 1957 that, unlike their American counterparts, Canadian entomologists had succeeded in maintaining a balance between economic entomology and the less practically oriented aspects of the discipline. He also noted that foreign entomologists believed that there was little interesting research done in the United States, meaning that the balance between academic and economic research had favored the latter. American entomologists, he concluded, admired their Canadian counterparts more than these foreigners because they had managed to combine the two.[101] Indeed, even as late as in the 1960s and '70s, when the major employer of Canadian entomologists, the Canadian Department of Entomology, was coming under increasing pressure to become more closely concerned with economic matters, they were still very unsure about the equation of their science with "pest management".[102] This attitude may have allowed Canadian entomologists, as has been claimed by the entomologist Edward Smith at Cornell University, to confront the polemics over the environmental impact of chemical insecticides such as DDT much better than their American counterparts.[103] Unlike the latter, Canadian entomologists had not staked their professional reputations so completely on the development of ever more effective methods of pest control. It seems significant that the only explicit criticism of Rachel Carson published in the presumably unsympathetic *Agricultural Institute Review* was by an American entomologist, Robert White Stevens from the American Cyanamid Corporation and Rutgers University.[104]

There were, however, major costs attached to this more even-handed orientation of Canadian entomology. The institutional and often geographical remoteness of the leading Canadian entomologists, especially of those concerned with the agricultural applications of the most recent advances entomology and ecology, from the world of agricultural production, meant that their expertise had little impact on the development of pest control programs. Farmers and the entomologists closest to them preferred chemical control to any other practice. By the 1950s they were thoroughly committed to the use of DDT and the new generation of very effective synthetic insecticides, and the destruction of salmon runs on the Miramichi river was witness to this. The dramatic split between the world of entomological research and practices of pest control, as we shall see, also meant that the response to these disasters by Canadian entomologists interested in more ecologically sophisticated alternatives to chemical control was very different from that of their American colleagues.

NOTES

1 For a discussion of natural historical studies in Canada during the early nineteenth century, see S. Zeller, *Inventing Canada: Early Victorian Science and the Idea of a Transcontinental Nation* (Toronto: University of Toronto Press, 1987), especially, pp. 215–7.

2 J.T.H. Connor, "Of butterfly nets and beetle bottles: the Entomological Society of Canada, 1863–1960", *HSTC Bulletin* 6 (1982): 151–71. On Canadian learned and scientific societies during the mid nineteenth-century, see P.J. Bowler, "The early development of scientific societies in Canada" in A. Oleson & S.C. Brown (eds*), The Pursuit of Knowledge in the Early American Republic* (Baltimore: Johns Hopkins University Press, 1976), pp. 326–39; and R.A. Jarrell, "The social functions of the scientific society in nineteenth-century Canada" in Jarrell & A.E. Roos (eds.), *Critical Issues in the History of Canadian Science, Technology and Medicine* (Ottawa: HSTC Publications, 1983), pp. 31–44.

3 P.W. Riegert, *From Arsenic to DDT: A History of Entomology in Western Canada* (Toronto: Toronto University Press, 1980), pp. 41–55.

4 C. Berger, *Science, God, and Nature in Victorian Canada* (Toronto: University of Toronto Press, 1982), pp. 31–50.

5 C. Sorensen, "Brethren of the net: American entomology, 1840–1880", (Ph.D. dissertation, University of California, Davis, 1984), pp. 275–309 and 361–3.

6 Quoted in G.J. Spencer, "A century of entomology in Canada", *Canadian Entomologist* 96 (1964): 33–59, on p. 35.

7 On the efforts of the Agricultural and Arts Association of Ontario to promote agricultural science during the 1860s and 1870s, see D.A. Lawr, "Agricultural education in nineteenth-century Ontario: an idea in search of an institution", *History of Education Quarterly* 12 (1972): 334–57.

8 Connor, "Butterfly nets and beetle bottles", pp. 153–9.

9 G.P. Holland, "Entomology in Canada" in G.B. Wiggins (ed.), *Centennial of Entomology in Canada, 1863–1963: A Tribute to Edmund M. Walker* (Toronto: Toronto University Press, 1963), pp. 7–13.

10 J. Fletcher, "Annual address of the president", *Annual Report of the Entomological Society of Ontario* 20 (1889): 2–8.

11 L.O. Howard, *A History of Applied Entomology* (Washington: Smithsonian Institution, 1930), p. 185.

12 V.C. Fowke, *Canadian Agricultural Policy: The Historical Pattern* (Toronto: Toronto University Press, 1946), pp. 29–110.

13 Fowke, *Canadian Agricultural Policy*, pp. 111–42.

14 L. Finlay & D.N. Sprague, *The Structure of Canadian History* (Scarborough: Prentice-Hall, 1979), pp. 157–60 and 195–201. See also W.L. Morton, *The Critical Years: The Union of British North America, 1857–1873* (Toronto: McClelland & Stewart, 1964); P.B. Waite, *Canada, 1874–1896* (Toronto:

McLelland & Stewart, 1971); and R.G. Brown & R. Cook, *Canada, 1896–1921: A Nation Transformed* (Toronto: McClelland & Stewart, 1974).

15 Fowke, *Canadian Agricultural Policy*, pp. 85–90 and 160–87.

16 L.B. Lee, "The Canadian-American irrigation frontier, 1884–1914", *Agricultural History* 40 (1966): 271–83; and J.M. Richtik, "The policy framework for settling the Canadian west, 1870–1880", *Agricultural History* 49 (1975): 613–28.

17 Fowke, *Canadian Agricultural Policy*, pp. 271–81; see also D.H. Breen, *The Canadian Prairie West and the Ranching Frontier, 1874–1924* (Toronto: University of Toronto Press, 1983), pp. 1–22.

18 See R.W. Winks, *The Relevance of Canadian History: US and Imperial Perspectives* (Toronto: MacMillan, 1979); as well as the essays by D.W. Noble in R. Kroes (ed.), *The American West as Seen by Europeans and Americans* (Amsterdam: Free University Press, 1989), pp. 19–36, C.A. Milner and K.L. Bryant in Milner *et al.* (eds), *The Oxford History of the American West* (Oxford: Oxford University Press, 1994), pp. 155–94 and 195–236.

19 Fowke, *Canadian Agricultural Policy*, pp. 137–40.

20 Riegert, *Arsenic to DDT*, pp. 35–36 and 67–68.

21 Holland, "Entomology in Canada", p. 11. See also V. DeVecchi, "The dawning of a national scientific community in Canada, 1878–1896", *Scientia Canadensis* 8 (1984): 32–58; "Science and scientists in government, 1878–1896 — Part 1", *Scientia Canadensis* 8 (1984): 114–42; and "Science and scientists in government, 1878–1896 — Part 2", *Scientia Canadensis* 9 (1985): 97–113.

22 R. Glen, "Entomology in Canada up to 1956: a review of developments and accomplishments", *Canadian Entomologist* 88 (1956): 290–371, on p. 294; and Howard, *History*, pp. 184–5.

23 See T.H. Anstey, *One Hundred Harvests: Research Branch, Agriculture Canada, 1886–1986* (Ottawa: Agriculture Canada, 1986), pp. 3–9.

24 Parliament of the Dominion of Canada, Sessional Papers (1886), Vol. VII, Paper no. 10: Report of the Minister of Agriculture for the Dominion of Canada for the Calendar Year 1885: Appendix 54: W. Saunders, "Agricultural colleges and experimental stations with suggestions relating to experimental agriculture in Canada", pp. 211–90, especially pp. 211 and 268.

25 Finlay & Sprague, *Canadian History*, p. 170.

26 See DeVecchi, "Dawning", p. 101; and I.M. Stewart, "Weld, William", *Dictionary of Canadian Biography*, Vol. XII, pp. 1093–4.

27 Fowke, *Canadian Agricultural Policy*, pp. 237–8.

28 Weld quoted in DeVecchi, "Dawning", p. 101.

29 Anstey, *One Hundred Harvests*, pp. 33 and 38.

30 See DeVecchi, "Science — Part 2", pp. 105–8; and F.T. Lord, "The development of agricultural entomology in Nova Scotia", *Proceedings of the*

Nova Scotia Institute of Science 33 (1986): 19–46, on p. 33; and W. Downes, "Fifty years of entomology on Vancouver Island", *Proceedings of the Entomological Society of British Columbia* 48 (1951): 9–16, on p. 16.

31 Riegert, *Arsenic to DDT*, pp. 72–80.

32 National Archives of Canada (Ottawa): record group 17: volume 3039, folder 40–2–1, Destructive Insect and Pest Advisory Board, 24 January 1928; see also Howard, *History*, p. 122.

33 DeVecchi, "Science — Part 2", p. 105.

34 On Canadian progressive reformism, see R. Cook, *The Regenerators: Social Criticism in Late Victorian Canada* (Toronto: Toronto University Press, 1985); and M. Valverde, *The Age of Light, Soap and Water: Moral Reform in English Canada, 1885–1925* (Toronto: Toronto University Press, 1991).

35 See Fowke, *Canadian Agricultural Policy*, pp. 246–7; and A. McLaren, *Our Own Master Race: Eugenics in Canada, 1885–1945* (Toronto: McClelland & Stewart, 1990).

36 W. Lochhead, "The teaching of entomology in our agricultural colleges", *Annual Report of the Entomological Society of Ontario* 43 (1912): 38–40, on p. 38. On Lochhead's involvement in the Canadian eugenic movement, see McLaren, *Our Own Master Race*, p. 13.

37 DeVecchi, "Science — Part 2", pp. 97 and 109.

38 B.G. Doern, *Science and Politics in Canada* (Montreal: McGill University Press, 1972), p. 27.

39 Anstey, *One Hundred Harvests*, pp. 23–25; see also C.G. Hewitt, "Applied entomology: its rise and progress", *Annual Report of Entomological Society of Ontario* 45 (1914): 29–41.

40 Doern, *Science and Politics*, p. 27.

41 Library of the Canadian Department of Agriculture (Ottawa): Gibson, *Entomology in Canada, with Special reference to Developments in the Dominion Department of Agriculture*, unpublished manuscript (1951), p. 15.

42 Anstey, *One Hundred Harvests*, p. 51.

43 Spencer, "Century", pp. 39–40.

44 C.G. Hewitt, "Some observations on the practical importance of the study of parasitic insects", *Annual Report of the Entomological Society of Ontario* 41 (1910): 62–4, on p. 62.

45 Quoted in Spencer, "Century", p. 39.

46 Quoted in Downes, "Fifty years of entomology", p. 11.

47 R. Bothwell *et al.*, *Canada, 1900–1945* (Toronto: Toronto University Press, 1987), pp. 33–4; and R.S. Harris, *A History of Higher Education in Canada*, 1663–1960 (Toronto: University of Toronto Press, 1976), pp. 169–71 and 275–82.

48 Lochhead, "The teaching of entomology in our agricultural colleges"; see also A.W. Baker, "The history and present status of entomology in the universities and colleges of Canada", *Annual Report of the Entomological Society of Ontario* 60 (1929): 33–46.

49 Harris, *Higher Education in Canada*, pp. 307–9; on research in Canadian universities, see H.B. Neatby, "The gospel of research: the transformation of English-Canadian universities", *Proceedings and Transactions of the Royal Society of Canada* 20 (1982): 275–84; and cf. Y. Gingras, "The institutionalization of scientific research in Canadian universities: the case of physics", *Canadian History Review* 57 (1986): 181–94.

50 DeVecchi, "Dawning", p. 50.

51 Howard, *History*, p. 185.

52 On the education of Canadian agricultural scientists, see Riegert, *Arsenic to DDT*, p. 211.

53 Anstey, *One Hundred Harvests*, pp. 348–51.

54 Spencer, "Century", pp. 41–2.

55 Gibson, *Entomology in Canada*, p. 93.

56 A. Gibson, "The development of applied entomology in Canada, 1914–1923", *Report of the Quebec Society for the Protection of Plants* (1924): 24–46, on pp. 37–8.

57 E.P. Venables, "Fifty years of entomology in the interior of British Columbia", *Proceedings of the Entomological Society of British Columbia* 48 (1951): 19–22, on p. 20.

58 Riegert, *Arsenic to DDT*, pp. 72 and 173–5.

59 National Archives of Canada: rec. group 17: vol. 3049: fold. 40–18 (2): K.W. Neatby to J.G. Taggart, 18 January 1951.

60 Doern, *Science and Politics*.

61 N.T. Gridgeman, *Biological Sciences at the National Research Council of Canada: The Early Years to 1952* (Waterloo: Wilfrid Laurier University Press, 1979), pp. 30–40.

62 See P. Henson, "Evolution and taxonomy: J.H. Comstock's research school in evolutionary entomology at Cornell University, 1874–1930", (Ph.D. dissertation, University of Maryland, 1990), p. 361.

63 Anstey, *One Hundred Harvests*, p. 59.

64 Anstey, *One Hundred Harvests*, pp. 61–2.

65 Gibson, *Entomology in Canada*, p. 94.

66 Glen, "Entomology in Canada", p. 335; and J.F. Hockey, *Agricultural Research in the Annapolis Valley Area, 1909–1965* (Ottawa: Queen's Printer 1967), p. 35.

67 M.C. King, *E.W.R. Steacie and Science in Canada* (Toronto: Toronto University Press, 1989). See also Bothwell *et al.*, *Canada*, pp. 339–340; and R. Bothwell *et al.*, *Canada since 1945: Power, Politics, and Provincialism* (Toronto: University of Toronto Press, 1981), p. 103.

68 Doern, *Science and Politics*, pp. 37–49.

69 National Archives of Canada: rec. group 17: vol. 3047: fold. 40–6(2); and vol. 3046: fold. 40–5–70.

70 Finlay & Sprague, *Canadian History*, p. 271.

71 Anstey, *One Hundred Harvests*, pp. 74–75; and Glen, "Entomology in Canada", pp. 350–63.

72 Anstey, *One Hundred Harvests*, pp. 86–93; on the British model, see T. DeJager, "Pure science and practical interests: the origins of the Agricultural Research Council, 1930–1937", *Minerva* 31 (1993): 129–50.
73 Glen, "Century", p. 298.
74 Glen, "Century", p. 291.
75 These estimates were reached by combining data in D.P. Jones, "Agricultural entomology" in R.F. Smith *et al.* (eds), *History of Entomology* (Palo Alto: Annual Reviews, 1972), pp. 307–32; and B.J. Benz *et al.*, "Publications of the Entomological Society of America: a descriptive study of patterns", *Bulletin of the Entomological Society of America* 33(4) (1987): 230–8.
76 A.C. Hodson, "A historical perspective of articles published in the *Canadian Entomologist*, 1868–1984", *Bulletin of the Entomological Society of Canada* 17 (1985): 72 -3.
77 W.E. Beckel, "Entomology in the university curriculum in Canada", *Canadian Entomologist* 96 (1964): 250–8, on p. 254. See also B.N. Smallman, "Perspectives on insect control", *Canadian Entomologist* 96 (1964): 167–71, on p. 170; and, more generally, P.A. Larkin & W.J.D. Stephen, *From Formalin to Fortran* (Ottawa: Science Council of Canada, 1971), p. 23.
78 National Archives of Canada: vol. 3049: fold. 40–18 (2): K.W. Neatby to J.G. Taggart, 18 January 1951.
79 Glen, "Entomology in Canada", p. 300.
80 Larkin & Stephen, *Formalin to Fortran*, pp. 20–1.
81 Finlay & Sprague, *Canadian History*, pp. 306–18; and Doern, *Science and Politics*, pp. 93 and 143.
82 J.G. Glassco, Report of the Royal Commission on Government Organization (Ottawa: Queen's Printer, 1963), vol. IV, pp. 265–270.
83 Glen, "Organization of the Research Branch"; and J. Ackerman, "Agricultural research: the role of economics", *Agricultural Institute Review* 18 (1963): 15–9 and 39.
84 T.A. Johnson, "Weak spots in Canada's agricultural extension system", *Agricultural Institute Review* 9 (1954): 25–6, on p. 26.
85 R.E. Balch, "Symposium on entomological education", *Annual Report of the Entomological Society of Ontario* 79 (1948): 58–65; and J.G. Rempel, "Trends in Canadian entomology teaching", *Annual Report of the Entomological Society of Ontario* 89 (1958): 21–6.
86 Venables, "Fifty years of entomology", p. 21.
87 R. Carson, *Silent Spring* (Boston: Houghton-Mifflin, 1987), pp. 129–38; see also R.O. Brinkhurst & D.A. Chant, *This Good, Good Earth: Our Fight for Survival* (Toronto: MacMillan, 1971), pp. 58–74 and 119–20.
88 Glassco, *Report*, vol. IV, pp. 238–40.
89 Anstey, *One Hundred Harvests*, p. 97–100; and Bryan Beirne (Simon Fraser University), interview with Paolo Palladino, 7 July 1988.

90 Finlay & Sprague, *Canadian History*, pp. 326–33; see also Harris, *Higher Education in Canada*, pp. 275–82 and 405–6.
91 Anstey, *One Hundred Harvests*, p. 63.
92 J.D. Detweiler, "Entomology and the arts curriculum", *Annual Report of the Entomological Society of Ontario* 61 (1930): 35–8. On academic attitudes towards specialization in Canadian universities, see P. Axelrod, *Making a Middle Class: Student Life in English Canada during the Thirties* (Montreal: McGill-Queen's University Press, 1990), pp. 39–64.
93 G.A. Moore, "Presidential address", *Proceedings of the Entomological Society of Ontario* 76 (1945): 5–9, on p. 8.
94 See T.R. Dunlap, "Organization and wildlife preservation: the case of the Whooping Crane in North America", *Social Studies of Science* 21 (1991): 197–221, on p. 201.
95 Bothwell *et al.*, *Canada since 1945*, pp. 125, 166–70 and 257–8.
96 Rempel, "Trends".
97 Glen, "Entomology in Canada", p. 317; and Larkin & Stephen, *Formalin to Fortran*, p. 26.
98 See Beckel, "Entomology in the university curriculum", pp. 253–4; and J.G. Rempel, "Trends in Canadian entomology: teaching", *Proceedings of the Entomological Society of Ontario* 89 (1958): 21–6, on pp. 22–3. See also D.K.M. Kevan, "The department of entomology at McGill University: a history to 1978", *Notes of the Lyman Entomological Museum Research Laboratory* 5 (1979): 1–88, on p. 30.
99 "Review of the 1975 manpower study", *Bulletin of the Entomological Society of Canada* 13 (1981): 10–6.
100 Sorensen, *Brethren of the Net*, pp. 141–4.
101 N. Turner, "Economic and non-economic research", *Proceedings of the Entomological Society of Ontario* 88 (1957): 52–3.
102 See D.C. Eidt, "Correspondence: pest management = entomology", *Bulletin of the Entomological Society of Canada* 9 (1977): 7–8.
103 E.H. Smith, "The Entomological Society of America: the first hundred years, 1889–1989", *Bulletin of the Entomological Society of America* 35 (1989): 10–32, on pp. 22–3.
104 R. White Stevens, "Feeding an exploding population: the role of agricultural chemicals", *Agricultural Institute Review* 17 (1962): 11–3 and 46.

CHAPTER 3

Biological Control in the United States and Canada

THE BASES OF BIOLOGICAL CONTROL

In the previous two chapters we have seen what institutional factors shaped the general development of entomology in the United States and Canada, and what structural hurdles might be in the way of any entomologist hoping to reverse those problems caused by DDT so vividly described by Rachel Carson in *Silent Spring*. Now I wish to turn my attention to those American and Canadian entomologists who were actually involved in attempts at reversal, by examining more specifically the history of their preferred approach to crop protection, biological control.

The biological control of insect pests, and other economically significant biological enemies of the farmer, such as weeds, relies on manipulating these species' biological and physical environment to check their numerical growth, and sometimes even reduce their abundance to more tolerable levels. Although today this approach includes a variety of different tactics, during the nineteenth and early twentieth centuries, the introduction of natural enemies, predators and parasites, or their multiplication by captive breeding and release, were by far the preferred ones. Many of the insect species attacking American and Canadian farmers' crops were accidentally imported, exotic ones. The pervasive, popular notion that different animal populations were maintained in their proper, balanced relationships by the divinely ordained mechanisms of predation and parasitism, and that the problems confronting farmers and growers were the result of their interference with these checks by introducing exotic crop and insect species, made proposals to introduce natural enemies a most obvious and appropriate solution to problems with insect pests.

There were, of course, other ways of readjusting this balance of nature gone awry (and remedying the accompanying economic losses). Quarantine programs might have been particularly useful to stem the flow of exotic insect species, but throughout the nineteenth century their establishment was constantly opposed, especially in the United States, because they were believed to infringe upon the freedom of commerce. Alternatively, the New World could be made less hospitable to these exotic species. For example, the damage done by the Cotton Boll Weevil could be limited by harvesting only the earliest cotton bolls, and promptly removing remaining plant stalks by plowing over the field at the end of the growing season, which left the Weevil with no over-wintering refuge. Apple orchards could be protected against the Codling Moth by keeping them clean and spacing trees far apart. Such cultural practices were, however, very unpopular because they required farmers to both reduce their productive stocks and spend more time tending to them. Furthermore, since these practices often worked only if all growers in any particular area agreed to adopt them, some form of legal enforcement was needed; cultural control was therefore also known, more disparagingly, at the "government method". The necessary authority was very rarely granted in the United States.

While some farmers doubted the philosophy of nature underpinning these biological and cultural approaches, there were no real alternatives until the introduction of arsenical insecticides in the early 1870s. Chemical control soon thereafter became the approach most preferred by both farmers and entomologists. It did not require any coordination of neighboring farmers' activities, as each could protect himself or herself without having to care in the least bit about what the neighbors were doing. Work on biological control continued, nonetheless, in some American and Canadian institutions well into the 1950s. Understanding how this was possible is very important for explaining why American entomologists were so much more successful than their Canadian counterparts in developing during the 1950s and '60s an approach to crop protection which would integrate the benefits of chemical insecticides and biological control agents.

BIOLOGICAL CONTROL AND THE UNITED STATES DEPARTMENT OF AGRICULTURE

In 1865, the State Entomologist of Illinois, Benjamin Walsh, suggested that many of the insects species attacking the crops of American farmers were introduced from abroad. He reasoned that these exotic species caused as much damage as they did because, unlike indigenous ones, they lacked their usual complement of natural enemies. This situation, Walsh believed, could be remedied if the natural enemies of the pests were also introduced to America.[2] The United States Department of Agriculture was very interested in the idea, but Walsh's proposal, was not put to the test until the 1870s, when Walsh's close friend, the Federal Entomologist Charles Valentine Riley, began to introduce natural enemies of the Cabbage Worm, with some limited success.[3] Then in the mid-1880s Riley and his assistant, Leland Howard came to believe that the Cottony Cushion Scale, which at the time was threatening to destroy the citrus industry of California, provided an almost ideal target for the approach.[4] Prompted by the challenge to their professional status by local, amateur economic entomologists intending to introduce some natural enemies on their own, Riley sought quickly to identify the Scale, and to find where it persisted at economically insignificant densities. He reasoned that wherever the Scale's existence was most marginal, there also had to be a great abundance of natural enemies. The place turned out to be Australia, and Riley despatched his assistant Albert Koebele to obtain from there as many of these natural enemies as possible. In 1888 Koebele returned from his voyage with just the Vedalia Beetle, which he then released in the citrus groves of southern California. Much to everyone's admiration, the Cottony Cushion Scale promptly, and almost permanently, disappeared.

Buoyed by the brilliant success in California, work on biological control was rapidly expanded. In 1905, Howard, Riley's successor at the Bureau of Entomology, established the Gypsy Moth Parasite Laboratory in Melrose, Massachusetts. This was the first institution in North America intended specifically to examine and advance the use of natural enemies as pest control agents.[5] Howard appointed his assistant, William Fiske, to direct the laboratory. Over the next five years Fiske made Melrose into a major center for research on the systematics and ecology of the Gypsy and Brown Tail Moths. Significantly, while engaged in this endeavor to control the two moth species, Fiske studied very closely the relationships between the populations of moths and those of introduced natural enemies, seeking to assess the relative importance of biological and climatic factors in determining the abundance of the latter populations.[6] Many American and Canadian entomologists who played an important role in the development of biological control, in North America and elsewhere, such as Harry Smith, Robin Thompson, and James Tothill, passed through the laboratory, and were profoundly inspired by Fiske's approach. The Bureau of Entomology mounted other major efforts in biological control by establishing, for example, the European Corn Borer Laboratory in Hyères, France.[7] All these efforts were constrained, however, by the framework of mission-oriented research promoted by Riley and Howard. That is to say, the various teams brought together for each project were disbanded and dispersed at their closure, and therefore no permanent institution dedicated to the development of biological control generally was ever created in the United States Department of Agriculture.

Over the long term, work in biological control fared far better in Hawaii and California, where the phenomenal success of the effort against Cottony Cushion Scale caught the imagination of entomologists and other officials in the State Board of Horticulture for California and the Hawaiian Sugar Planters' Association. Hawaiian sugarcane growers thought that the same approach might help their industry to fend off destruction by the exotic Sugarcane Leafhopper. The Leafhopper seemed to be completely unaffected by any existing chemical insecticides.[8] In 1893, therefore, these growers convinced the Territorial

Government of Hawaii to offer to Koebele the position of Territorial Entomologist. Koebele left his post as agent of United States Department of Agriculture to take up the offer. During the next ten years, he and other entomologists began an intensive program of foreign exploration and introduction of natural enemies to protect Hawaiian sugarcane plantations. It seems, however, that during this time the United States Department of Agriculture (and the federal government, generally) became increasingly unhappy with the political power that the sugarcane plantation owners wielded in the Territory, and began to withdraw its support for work on sugarcane, while encouraging production of other crops, like the pineapple. In 1904, the Hawaiian Sugar Planters' Association decided to establish its own entomology department, in the private experiment station which it had been supporting since 1895. Over the next forty years, this entomological department continued the work begun by Koebele with some notable successes, such as the control of the Leafhopper, but the use chemical insecticides eventually became the preferred approach to protecting the local sugar plantations.[9]

The most enduring American program in biological control was that established in California. The State Board of Horticulture there had provided much support for Koebele's work on the Cottony Cushion Scale, and was so excited by its outcome that it sought to expand the program of exploration and introduction. The Bureau of Entomology of the United States Department of Agriculture, embroiled in bitter disputes with the Board over who was responsible for the introduction of the Vedalia Beetle, opposed to this expansion.[10] Federal entomologists viewed the program of the Board of Horticulture as reckless, fearing that it would result in the accidental introduction of more undesirable exotic species. They argued that careful and extension taxonomic work was needed to identify the pests and their natural enemies, and to make sure thereby that any introduced species were in fact what they were thought to be, beneficial rather than noxious. Given this difficulty, and the expense of travel abroad to locate the natural enemies, they claimed that this work should not be left in the hands of amateurs, such as those working for Board of Horticulture, but should instead rely on the skills of professionally trained entomologists like those in Bureau of Entomology of the United States Department of Agriculture.[11] The Board, supported by powerful agricultural associations, proceeded nonetheless to establish a State Insectary, where the staff would multiply the predators and parasites it imported from abroad, often running afoul of federal regulations controlling the movement of insects. This tension between the federal and state organizations was lessened, however, during the first years of the century, as "progressive" reforms of Californian politics led to the appointment of a senior scientist rather than the usual political figure as Commissioner of Horticulture. In 1913, this new Commissioner, the former professor of entomology at Michigan Agricultural College, Albert Cook, appointed in turn a professionally trained entomologist, Harry Smith, to run the State Insectary.[12] The antagonism between the federal and state authority was to remain, nonetheless, a major problem for entomologists in California long thereafter.

BIOLOGICAL CONTROL IN THE CITRUS EXPERIMENT STATION

Harry Smith was one of the many graduates of the federal Gypsy Moth Parasite Laboratory at Melrose, and was especially well regarded by its director, William Fiske. At first, his appointment to the State Insectary in California was opposed by Smith's erstwhile employer, Leland Howard, but he reassured the latter by promising to bring the Californian program in line with the more cautious policies of the Bureau of Entomology.

Between 1913 and 1923, Smith coordinated an extensive program of foreign exploration, importing, multiplying, and introducing predators and parasites of insect pests attacking all sorts of crops grown by Californian farmers, and the citrus crop in particular. Although his program was more technically sophisticated than the earlier one, Smith's efforts continued nonetheless to attract much criticism from a number of different quarters. During the 1890s, agricultural scientists in the University of California had promoted the use of chemical insecticides as the most "progressive" way to control the Cottony Cushion Scale, but lost out to the staff directly employed by the state and federal government.[13] This situation only worsened the already precarious position occupied by these academic scientists, and some of them began therefore to question the scientific, more specifically, ecological, foundations of biological control.[14] Proponents of chemical control outside the university were equally critical, charging that Smith's program was not practical. While this latter criticism was particularly threatening because the stability of Smith's position was strongly tied to promises of economic achievements, Smith was also very sensitive to the criticism by the academic scientists because, ever since his days in the Bureau of Entomology, he had hoped eventually to obtain a post in a university.[15] The answer to his problems came from citrus industry of southern California.

The spectacular success of the Vedalia Beetle against the Cottony Cushion Scale made citrus growers great supporters of biological control; chemical insecticides, such as hydrogen cyanide, had not been particularly effective against this and many other insects attacking the citrus crop, and cultural practices to control them were unacceptable as they required too much labor and sacrificing productivity. This last point was well appreciated by Smith, who, while still working for the federal Bureau of Entomology, had witnessed the reluctance of cotton growers in the South to adopt cultural modifications to control the Cotton Boll Weevil.[16] Citrus growers were also very "progressive"; in 1905 they had organized themselves into a cooperative, the California Fruit Growers' Exchange (better known today as Sunkist), which soon thereafter mounted a powerful and very effective campaign to establish an agricultural experiment station in southern California focusing exclusively on the problems of the citrus industry.[17] In the same year, the State Legislature established in Riverside, in the heartland of the citrus industry, a Citrus Experiment Station; Herbert Webber, a geneticist and plant breeder from the Bureau of Plant Industry of the United States Department of Agriculture, was appointed as its first director. Entomologists recruited into this new station were as skeptical about biological control as their counterparts in the University of California, but this did not matter to the members of the Fruit Growers' Exchange. In 1923, the staff of the State Insectary was transferred to the Citrus Experiment Station, and Smith was appointed director of a new Division of Biological Control, separate from that for entomology. The historian Richard Sawyer has argued that this institutional separation was particularly important for the subsequent growth of Smith's program; the program in biological control in the federal Bureau of Entomology declined because, he claims, it was never granted such a separate institutional base.[18]

Soon after the removal to his new, more secure home in Riverside, Smith began to take up the questions raised by his critics in the University of California about the ecological foundations of biological control. He focused, in particular, on the effectiveness of multiple introductions of different predatory and parasitic species to control any one pest, an approach whose validity, he believed, was confirmed by the second great success of biological control in California, the control of the Citrus Mealybug by the *Cryptolaemus* beetle.[19] For Smith, the action of natural enemies was clearly and fundamentally important to maintain insect populations at their proper level. He extended the theoretical lines of work started by Fiske, his mentor in the Gypsy Moth Parasite Laboratory at Melrose, to criticize all those who questioned this equilibrating role of predation and parasitism. His belief in this equilibrium was totally unaffected by some of these critics' observations that the abundance of *Cryptolaemus* did not always follow that of the Mealybug. As these critics saw it, the winter climate often reduced the abundance of both species, and in the spring it took the beetle some time to build up to densities at which they had an impact on the populations of the Mealybug. This meant that each spring citrus groves would have to be inundated with beetles which had been raised during the winter in the insectaries on the Citrus Experiment Station or in those operated by private businesses.[20] In other words, the approach was no different from the use of any chemical insecticides, and the relative costs of each approach had to be judged on that basis, not on some spurious claim to permanent control achievable only by biological means. Sawyer has suggested that a common assumption among economic entomologists that, if insect pests caused no damage, as happened with the Mealybug during the winter, they were uninteresting, explains Smith's lack of attention to what exactly happened to *Cryptolaemus* in the winter.[21]

During the 1930s, this situation began to change as the entomologists in the Division of Biological Control became less concerned about any immediate, practical problems. As Robert Vickery of the California Spray Chemical Corporation admitted in 1935, the institutional situation of biological control was much improved as the Great Depression and the relative cheapness of this approach made it an increasingly attractive option for distressed farmers.[22] Smith became progressively less involved with the daily work of insect identification, raising, and multiplication, and much more concerned with the empirical foundations of mathematical theories about the dynamics of insect populations.

Ecologists from all around the world became very interested in his work in this area. One of the most important proponents of theoretical population ecology, John Nicholson, traveled from Australia to visit the Station a number of times, and so did the British ecologist George Varley, who spent a whole year at the Station as a postdoctoral student.[23] Smith was not the only entomologist in the Division to attract visitors. Many came to learn about esoteric biological problems from insect systematists and ecologists such as Stanley Flanders and Paul DeBach. These two had obtained their doctorates from the University of California under Smith's supervision during the 1930s, and were then hired by the Citrus Experiment Station. DeBach and Smith soon co-authored a paper on experimental testing of Nicholson's ideas which was then published in *Ecology*, the leading journal in the field. Interestingly, limited their tests to the first seven generations of interactions between the domestic houseflies and their pupal parasite, *Mormoniella vitripennis*.[24] Although they had hoped to strengthen their results by extending these tests over a greater number of generations at a later date, they never did. The reasons for this are unclear, but it seems likely that a six months-long study of houseflies would have been viewed with great suspicion by his patrons and would have been soon discouraged. Smith was always very careful that very little about his work on the dynamics of insect populations should reach the ears of these patrons; with them, he continued to cultivate an image of biological control work as a staid and empirical practice.[25]

Smith's fame also grew among American economic entomologists, who, in 1941, elected him president of the American Association of Economic Entomologists. In his inaugural address, he took up his colleagues' increasingly exclusive interest in chemical insecticides.[26] He warned them that it was not sufficient to study just the physiology of insects to develop better chemical control programs. The ecological dynamics and genetics of insect populations had also to be taken into account to understand, for example, some insects' resistance to insecticides. Significantly, this view was informed by the most recent developments in population genetics; when Smith's group in the Division of Biological Control became interested in the development of resistance, they turned for help to their neighbors in the California Institute of Technology, and to Theodozius Dobzhansky in particular.[27] Even more significantly, Smith argued that the answer to the problem was not to switch to biological control, but to alter chemical control programs in light of both these recent developments in genetics and the findings of workers in biological control. For all his increasing interest in the theory of population ecology, Smith's ultimate commitment was still to improve methods of crop protection.

BIOLOGICAL CONTROL AND DDT

Chemical insecticides had increased in popularity so steadily during the 1920s and '30s that biological control work had been virtually abandoned everywhere in the United States, except for the large program in the Citrus Experiment

Station and the smaller ones supported by the United States Department of Agriculture and the Hawaiian Sugar Planters' Association. While the Great Depression had granted these programs a new lease on life, the latter two were conclusively swept aside, soon after the second world war by the introduction of DDT and related synthetic insecticides.[28]

Entomologists in the University of California and the Citrus Experiment Station reacted to this new situation by arguing that it was imperative to consider the ecological and evolutionary implications of DDT before using it as blindly as everyone seemed to be doing. Abraham Michelbacher, from the Department of Entomology at Berkeley, argued in an editorial in the *Journal of Economic Entomology* that, although nearly every entomologist recognized the importance of ecological work in the development of pest control programs, many had to handle so many other problems that they had no time to study insect ecology.[29] This situation, Michelbacher believed, was directly responsible for the excessive use of chemical insecticides, and the consequent poisoning of the soil, destruction of natural enemies, and pests' increasing resistance, which had stymied the apple industry in the American Northwest during the late 1930s. Without greater support for entomologists' ecological work, the use of DDT would soon lead to similar problems. Such arguments carried very little weight until the Cottony Cushion Scale became, once again, a major problem for the citrus industry of southern California.

In 1946, heavy use of DDT over seventy-five percent of the citrus groves in the San Joaquin Valley, to control an outbreak of the Citricola Scale, resulted in the heaviest infestations of the Cottony Cushion Scale since the introduction of the Vedalia Beetle in the 1880s. Smith's ever pragmatic response to this situation was to explain to the growers that the Vedalia Beetle was far more susceptible to DDT than either the Citricola or Cottony Cushion Scales, and that its controlling effect over the abundance of the latter two species had therefore been destroyed. He convinced them to halt the use of DDT for one year, which soon allowed the Vedalia Beetle to re-establish its control over the Cottony Cushion Scale.[30] The power of natural enemies was confirmed once again, and the citrus industry and state legislature renewed their support for the work of the Division of Biological Control. This allowed for some limited institutional expansion, as the brief of responsibilities for the Citrus Experiment Station came to include insects affecting growers in northern California, and a substation of the Division of Biological Control was established in Berkeley to aid this new work.[31]

Smith's former student, Paul DeBach, took advantage of these ecological problems with DDT to raise the profile of biological control in the American Association of Economic Entomologists, and in 1949 he succeeded in establishing, within the Association, a special section dedicated to biological control.[32] Nevertheless, even in California, funding for work on biological control did not keep pace with that for chemical control, and the number of posts for biological control specialists did not grow anywhere nearly as fast. Most entomologists acknowledged the problems with DDT, but a growing, more sensible use of DDT, and the emergence of newer, more powerful alternatives, such as organophosphates like malathion, reassured them that they were on the winning track. Smith was soon discouraging students from working on biological control because he did not see any employment opportunities for them in the future. Neither did he try to find alternative sources of funding to keep work in this area going, simply assuming that if alternatives to biological control did the job properly, as he thought the new generation of insecticides could, then there was no need for more work in biological control.[33] If these alternatives did not, in fact, do the job, then more funds would be provided for his and his colleagues' work. That is how it had always worked during his career, which came to an end with his retirement from the Division of Biological Control in 1951. He was succeeded by Curtis Clausen, who had just left the United States Department of Agriculture, disgusted by the lack of support there for biological control.

Even though, on his retirement, Smith left behind him the only institutionally independent program in biological control in the United States, those staffing it realized that the future of the Division of Biological Control did not look promising. They agreed that some new arrangements had to be found, but were deeply divided about what form these arrangements should take.

INTEGRATED CONTROL

It was during this difficult period for biological control that the entomologists in the Division of Biological Control began to articulate their solution to the increasingly obvious problems caused by the unrelenting commitment to chemical control. It would be a chemical control program that caused the least possible disruption of natural control mechanisms, and was compatible with introduced biological agents. However, it was obvious to these entomologists that the broad spectrum of action of insecticides such as DDT precluded intervention against any one species, and that general destruction of natural enemies was therefore unavoidable. Furthermore, the polarization of the American entomologists usually prevented any such constructive interaction between the proponents of chemical and biological control. For example, in the University of California at Berkeley and Riverside there were two separate departments of entomology and biological control on each campus, and the relationship between them was not amicable.[34] However, problems confronting the dairy industry of California during the 1950s, and external funding to confront them from a very unusual source, provided some entomologists in the divided departments at Berkeley and Riverside with an opportunity to build their alliance of chemical and biological control.

Alfalfa was the mainstay of the dairy industry, and its production, in California, was seriously threatened by the sudden appearance of the Spotted Alfalfa Aphid. The problem facing the industry was how to protect their productivity without running afoul of the Food and Drug Administration and the Public

Health Service. Milk had been one of the agricultural products they most carefully scrutinized for insecticidal residues, and early on they had set a "zero tolerance" for DDT in milk.[35] DDT moved too easily from the alfalfa, through the cow, into the milk. The very toxic, but much more labile organophosphate insecticides, such as malathion and parathion, however, left no residues, and thus could be used against the Alfalfa Aphid. Unfortunately, these insecticides destroyed the Aphid's natural enemies to a far greater extent than DDT, and in some places the Aphid was becoming resistant to them.[36] Control was increasingly difficult. Alfalfa growers were in a quandary.

Ray Smith, Abraham Michelbacher's former student and member of faculty in the Department of Entomology at Berkeley, responded to this situation by initiating a system of "supervised control", similar to that established during the 1930s by Allison Pickett in Nova Scotia. He placed students with alfalfa growers to help the latter to time their uses of insecticides so as to minimize the damage done to natural enemies of the targeted insect species.[37] Soon thereafter, Smith and other entomologists in the departments of entomology and biological control at Riverside and Berkeley, helped by unprecedented funding from the National Institutes of Health, began to articulate a more systematic and general answer to the problem confronting alfalfa growers and the dairy industry.[38]

Vernon Stern and Smith, from the departments of entomology at Riverside and Berkeley, along with Ken Hagen and Robert Van den Bosch, two of Smith's former students and early participants in the supervised control scheme, then employed in the Divisions of Biological Control, combined their knowledge of chemical insecticides and biological control to search for an insecticide whose use would not lead to the same problems caused by malathion and parathion. This approach had already been tried with some success by a number of entomologists, especially by some of Harry Smith's students at Riverside.[39] Stern and his collaborators soon found that parasites of the Alfalfa Aphid were resistant to Systox when they were in the pupal stage; furthermore, since its breakdown was relatively rapid, Systox would have minimal impact on the complex of natural enemies. A high degree of control could be obtained with a minimal use of insecticide. Another crucial, and more radical element of this "integrated control" approach was the idea developed in the course of supervised control programs that insecticides should be used only when the abundance of insect species attacking the crop overtook the appropriate "economic threshold".

Since the turn of the century, and even more so after the advent of DDT, insecticides tended to be used on fixed schedules, determined by spray calendars, as a form of insurance against insect outbreaks. Since it was very difficult to predict whether any insect population was going to remain at the observed level, or whether it was going to collapse naturally, or whether it was going to explode and reach a destructive level, the best strategy seemed to be to destroy any apparent infestation. Moreover, since the value of the crop saved each year easily outweighed the outlay for insecticides, it seemed wise to use them on a fixed schedule. Economic thresholds were meant to break this tradition by stipulating that control measures should be adopted only when the projected economic losses incurred by the growers exceeded the costs of insecticidal treatment. The definition of the costs was, furthermore, broader than in its traditional conception because the value of the crop saved following the insecticidal treatment was supposed to be weighed against the damage done to the natural enemies of the targeted species, and the concomitant increased cost of protection. Under this broader set of considerations, the appeal of insecticide treatment was severely dampened; it was conceivable, for example, that in some cases no action was the economically rational response even in the presence of pest outbreaks.[40]

In some ways, however, integrated control, as defined by Stern and his colleagues, did not represent a radical departure from prevailing entomological norms. The approach was undoubtedly an ecologically informed one because its development was based on an unusual awareness of the importance of natural enemies and of life history in the control of insect pests. Yet, the success of the program actually rested on the rather unique features of Systox; in fact, its proponents acknowledged that it would have been very difficult to build the approach with a common, broad spectrum insecticide, because in such a situation a much more "intimate knowledge of the behavior patterns of the pests and their natural enemies is required".[41] But there were very few selective insecticides other than Systox, and the incentives for the chemical industry to produce more of them were equally few. In their absence, the future of integrated control would have to rest on the development of this more intimate knowledge of insect ecology, a direction in which some of the leading entomologists in the Division of Biological Control were well prepared to go.

FROM BIOLOGICAL CONTROL TO ECOLOGICAL RESEARCH

Harry Smith, while still the director of the Division of Biological Control in the Citrus Experiment Station, had granted relatively wide discretion to those who worked under him over the conduct of their particular projects, but he had also made sure that his patrons were always well served. He ignored, for example, federal opposition to his growing involvement in the biological control of weeds, which he began in response to calls from strawberry growers in northern California to do something about the Klamath Weed, and recruited a young insect ecologist from Ohio State University, Carl Huffaker, to develop a program against this weed. However, Huffaker, and other younger researchers in biological control brought in by Smith during this period, had very different ideas about how to tackle such problems.

By the 1950s, advancing understanding of the scientific foundations of agriculture was becoming in the University of California (as in many other centers for agricultural research and education) far more important than providing any direct service to agricultural producers.[42] Thus, for Smith's more ecologically inclined successors — Huffaker's first degree was in plant ecology, and he became interested in insects only after his doctoral studies in biological control under Alvah Peterson, at Ohio State University — experimental studies of predation and parasitism, grounded in the most recent advances in theoretical ecology, were crucial prerequisites for the development of more effective programs of biological control.[43] Funding was now available from the agricultural experiment station for Huffaker to begin a series of experiments that have now become part of ecologists' canonical list of classic studies on the population dynamics of predation and parasitism; Paul DeBach's contemporary work on competition between allied species of parasites for a common host is equally significant in the history of ecology.[44] Moreover, these entomologists' vision of biological control and its arsenal was far broader than Smith's. For them, cultural modifications, as well as the introduction of microbial agents such as *Bacillus thuringiensis,* which had been pioneered by their colleague, Edward Steinhaus, were as important as the introduction of parasites and predators, and none of these approaches, just like chemical control, made any sense if not viewed as elements of a much broader ecological approach to crop protection. During the mid-1950s, they sought to advance this rather different understanding of their proper roles and goals by demanding the establishment in Berkeley of an administratively separate Division of Biological Control.

Not everyone in the Division of Biological Control in Riverside was, however, prepared to abandon Smith's ideas about the organization of biological control. For example, Charles Fleschner, the director of the Division from 1959 to 1963, sought to focus all research on biological control alone. He also opposed any co-operation with the Department of Entomology.[45] But, having emphasized so much their separate identity, and having tied this identity so closely to the needs of specific groups of agricultural producers, the position of older experts in biological control such as Fleschner became untenable as their institutional environment was fundamentally transformed in the postwar years. In 1963 Fleschner was replaced by Donald Chant, who was brought into the Division of Biological Control from the Canadian Department of Agriculture Research Laboratory in Vineland, Ontario, specifically to move the Division in the new direction being taken by the University of California. Chant, as we shall see in the next chapter, was particularly critical of what he saw as Smith and his colleagues' theoretically unsophisticated approach to biological and integrated control. His promotion of a more intense focus on the development of the latter strategy led eventually to the merger of the Division of Biological Control in Riverside with the Department of Entomology.

The history of biological control in Berkeley was quite different. In 1963, Huffaker and others finally succeeded in establishing there an independent Division of Biological Control, which has maintained its independent status to this day. This was because, Huffaker and Ray Smith brought about, in Berkeley, a much deeper institutional and intellectual involvement with ecological research, to transform integrated control during the 1960s and '70s into what the entomologist Marcos Kogan has called, polemically, a triumph of "applied ecology".[46] Before going on to examine this transformation, however, we need to study the development of biological control in Canada. The implicit contrast between American and Canadian work in this area will allow better understanding of the ambiguities surrounding Kogan's description of integrated control.

BIOLOGICAL CONTROL IN CANADA

Just as in the United States, biological control received a great deal of attention from Canadian governmental entomologists during the late nineteenth century. In the late 1880s, the first director of the Dominion Experimental Farms, William Saunders, introduced parasites to control insects damaging crops in Ontario. Saunders' colleague, founder, with Saunders, of the Entomological Society of Canada, and first Dominion Entomologist, James Fletcher, followed suit, with much encouragement from Leland Howard, at the United States Department of Agriculture.[47] Howard, like many other entomologists, was very hopeful about the future of biological control, and intended to extend the domain of his Bureau of Entomology through the exchanges of predators and parasites needed to establish this approach.

Biological control in Canada, however, received its strongest boost with the appointment of Gordon Hewitt as the second Dominion Entomologist in 1909. One consideration involved in his appointment was that forests in eastern Canada were at this time threatened by a major outbreak of the European Larch Sawfly, and, while still in Britain, Hewitt had worked on the natural history and biological control of this insect.[48] Hewitt continued this work in his new home, and began to build a research program around the Sawfly, by bringing back from the United States a student of John Henry Comstock at Cornell University, James Swaine, and setting him to work on the introduction of those parasites which could be found in Britain.[49] Hewitt was clearly committed to the development of biological control as an alternative to the ever more popular use of chemical insecticides.[50] He felt that entomologists spent so much time studying how to improve these insecticides that they both underestimated the possibilities of biological control, and lost sight of the great importance of insect ecology. He, however, planned to change how things were done in the Canadian Department of Agriculture.

In 1912, Hewitt took advantage of the arrival of the Brown Tail Moth in Nova Scotia and New Brunswick to force the establishment of the Entomology Laboratory at Fredericton, New Brunswick.[51] James Tothill, a graduate of Ontario Agricultural College and Cornell University, was called to direct the work of the Laboratory, and, more specifically, to extend the work of parasite and predator introductions which he had begun while working in the United States Department of Agriculture Gypsy Moth Parasite Laboratory at Melrose.[52]

In the same year, Hewitt also succeeded in establishing a Natural Control Investigations Laboratory in Royal Oak, British Columbia, which was to focus on the problems caused by the Spruce Budworm.[53]

By the early 1920s, the infestations of the Larch Sawfly and Brown Tail Moth had collapsed; similar collapses seemed to follow introductions against other insects affecting fruit crops.[54] These successes, and the increasingly acute problems caused by the seemingly unstoppable advance of the European Corn Borer in eastern Canada, served to justify the further expansion of support for biological control work. Moreover, reflecting growing national assertiveness, Hewitt's successor, Arthur Gibson, felt that the Entomology Branch was too dependent on Howard's Bureau of Entomology for the importing of parasites to control the Corn Borer and Brown Tail Moth in Canada, and that the branch should endeavor instead to have collections made by its own entomologists.[55] Most Canadian work in biological control was, therefore, first consolidated, and then expanded under Tothill's direct supervision; for example, the Natural Control Investigations Laboratory was removed from Royal Oak and brought to Fredericton, and a search was started in response to demands by the Ontario Corn Growers Association for a suitable site somewhere in Europe to establish a laboratory especially dedicated to the Corn Borer.[56] Then, in 1923, Tothill was appointed director of the newly established Forest Entomology Laboratory, also at Fredericton, and the Natural Control Investigations Laboratory was transferred to Chatham, Ontario, where the staff began to work, in the midst of the local corn fields, almost exclusively on the Corn Borer. In 1929, the Natural Control Investigations Laboratory was moved one last time to Belleville, Ontario, and was renamed the Dominion Parasite Laboratory.[57] This was to become the premier center for work on biological control, a distinction which it retained until 1970, when the Laboratory was disbanded and biological control ceased to be an institutionally distinct area of entomological work in Canada.

EXPANSION OF BIOLOGICAL CONTROL

Although biological control may have been a popular approach to protect forests from the European Larch Sawfly because using a chemical insecticide against it would have been prohibitively expensive, this approach was not so popular among some very important Canadian agricultural producers, the fruit growers in the Okanagan Valley of British Columbia. During the 1920s, they often expressed as much doubt about the merits of work on insect taxonomy and natural history as their American counterparts.[58] Such studies revealed to entomologists that most of the growers' problems arose from the accidental introduction of exotic insect species, and, as such, were amenable to biological control. Growers, however, were very much opposed to the proposed cultural modifications which also seemed to flow almost predictably from these investigations. They knew quite well that chemical insecticides were far more convenient and effective against, for example, the Codling Moth, than any number of introduced parasites and predators, or any form of cultural control. They preferred insecticides, and believed that entomologists should work to make them more effective still. The use of these insecticides therefore grew in Canada just as rapidly as it did in the United States.[59]

Notwithstanding this apparently adverse environment, and the failure to produce anything that was nearly as successful as the Vedalia Beetle was against the Cottony Cushion Scale, Canadian entomologists interested in biological control could take steps to increase support for their work that were simply inconceivable for their counterparts in the Citrus Experiment Station at Riverside. Although their introductions of parasites and predators were sometimes followed by reductions of the targeted insect populations, such as those of the Brown Tail Moth and the European Larch sawfly, there actually was very little evidence to support the view that the collapse was in fact due to those introductions. Extraordinarily, the director of the laboratory at Belleville, Alfred Baird, himself pointed out this problem, thus admitting indirectly that the effectiveness of biological control was limited, and attracting Harry Smith's criticism that Canadian entomologists were very conservative.[60] Baird, another graduate of Cornell University, responded to this situation by calling for greater support for research in taxonomy and systematics to understand how biological control worked because only this would improve its effectiveness.[61] Authorities in the Dominion Department of Agriculture appeared to agree with Baird, as they continued to support work in biological control.

By the mid-1930s, the Dominion Parasite Laboratory was staffed by more numerous and better trained entomologists than the Citrus Experiment Station at Riverside. Baird could count on eleven assistants, all of whom had received some post-graduate training in entomology. Furthermore, the physical facilities at Belleville were expanded considerably in 1936, and a number of new projects were started.[62] This expansion was celebrated by organizing a conference on "Biological Methods of Controlling Insect Pests", which attracted to Belleville leading ecologists and workers in biological control. Smith was there, as was Alvah Peterson from Ohio State University, who was then supervising Carl Huffaker's doctoral dissertations. Curtis Clausen from the Bureau of Entomology in Washington, and Royal Norton Chapman's former student at the University of Minnesota and then professor of zoology at Queen's University, the population ecologist John Stanley, also attended. Lastly, but not least, the eminent entomologist and population ecologist, Robin Thompson, came all the way from London, as representative for the Imperial Institute of Entomology. Interestingly, James Swaine, then director of the Science Service of the Dominion Department of Agriculture criticized on this occasion Stanley's laboratory studies on the population dynamics of the Flour Beetle for their scarce relevance to understanding the dynamics of insects populations in more natural settings. Swaine and Peterson were in turn criticized by Baird, Smith, and Thompson, who argued that effective biological control depended on holding onto the "economic viewpoint" which they seemed all too ready to abandon with their call for more studies of "non-economic" species.[63] These tensions over the relationship between biological control and ecological studies were, as we shall see in the next chapter, very important for the subsequent development of both biological and integrated control.

Work on biological control also received a strong boost from the Canadian lumber and pulp industries.[64] Protecting Canadian forests against the Spruce Budworm or the European Spruce Sawfly was very important to these industries. But, even in the 1930s and '40s, there were no effective chemical insecticides for the job; using arsenical insecticides over thousands of forested acres would have been far too difficult and costly.[65] These very important and influential industries encouraged, therefore, the expansion of biological control programs, and advancement of any other areas of research which might lead to improved control over the damage caused by, for example, the Spruce Sawfly.[66] Following the discovery of the Sawfly in the forests of eastern Quebec in 1931, the federal government increased its support for workers in biological control, and especially for their researches on the statistics and dynamics of insect populations, which the Canadian Pulp and Paper Association and the Canadian Lumbermen Association hoped would help to improve the forecasting of destructive outbreaks.[67] Arthur Gibson, who was at this time the director of the Entomology Branch, remarked in his history of entomology in Canada that this marked the beginning in Canada of "fundamental work" on the ecology of forest insects.[68] But it was the advent of war in 1939, and the accompanying, dramatically increased demand for lumber, which truly accelerated the development of these programs to a previously unimaginable extent.[69] Losses to insects became so important that the father of postwar reconstruction, "C.D." Howe, extended governmental involvement with problems caused by forest insects by providing, in the Reconstruction Act of 1944, for the organization of a Forest Insects Control Board, which immediately recommended the establishment of a Forest Insect Laboratory, in Sault Ste Marie, Ontario, for work on both biological and chemical methods of control.[70]

Thanks to the needs of the very important lumber and pulp industries, and the political distance between agricultural producers, such as those in the Okanagan Valley, and entomologists in the Department of Agriculture in Ottawa, the fortunes of biological control during the 1930s seem to have been much better in Canada than the United States.

CHANGING DIRECTIONS

Even though Canada, by the 1930s, had become quite independent from Great Britain, it did not escape the trials and tribulations of the Empire prior to this time, the work in biological control of the Dominion Parasite Laboratory was closely coordinated with that of the Imperial Bureau of Entomology in London.[71] Following the Imperial Agricultural Research Conference in 1927, which was intended to bring about greater coordination to the increasingly fragmenting imperial activities in the area, it was agreed that the primary responsibility for foreign explorations to discover natural enemies for introduction in Britain or in any of the colonies and dominions of the British Empire, should lie with the Imperial Bureau of Entomology, and in particular with Robin Thompson's Parasite Laboratory at Farnham Royal, near London. The principal responsibilities of the Canadian entomologists in the Dominion Parasite Laboratory were, then, to identify any insect species attacking a particular agricultural crop, and then turn the information over to the Imperial Bureau. The Bureau would then seek the appropriate natural enemies, and eventually ship them to the Dominion Parasite Laboratory, for multiplication, distribution and introduction against the pest in question. This connection was strengthened in 1940, when the Parasite Laboratory at Farnham Royal was closed, and Thompson and his staff were relocated in Belleville.

Although formally belonging to separate organizations, the entomologists newly arrived at Belleville from Great Britain and those employed by the Dominion Parasite Laboratory collaborated quite closely. Those who were most interested in the ecology of insect populations, such as Thomas Burnett, who had been Thomas Park's student at the University of Chicago and was to become a source of inspiration to an entire generation of Canadian population ecologists, often turned to Thompson for advice on the more esoteric questions of biological control.[72] This relationship, however, led many outsiders to view the Laboratory simply as part of the Imperial Bureau of Entomology. This perception was strengthened when Thompson became the editor of the *Canadian Entomologist*, and came to be regarded, therefore, as a more senior figure among Canadian entomologists than Alfred Baird. There was much confusion over the relationship between Thompson's organization and the Dominion Parasite Laboratory, which rather unsurprisingly irked Baird. This tension continued after the second world war. Paralleling the rapidly loosening ties with Great Britain, Canadian entomologists began to organize their own exploratory missions to Europe for parasites and predators. Interestingly, Thompson sought to protect his domain by arguing that European entomologists were better prepared for the work of exploration and collection because they were better trained than their Canadian counterparts in the "fundamental" aspects of entomology.[73] Finally, in 1948, Thompson's and Baird's organizations, now renamed as the Commonwealth Institute for Biological Control and the Biological Control Investigations Unit of the Canadian Department of Agriculture, were geographically separated by removing the former to Ottawa. Like many matters in the Commonwealth, the boundaries between the two institutions remained, nonetheless, very confused in these rapidly changing times.

This sense of confusion was aggravated as the old exclusive emphasis on the implementation of biological control programs, which had been strongly promoted by Baird, continued to be stressed by his more academically inclined, but equally old-fashioned, successor, the systematist Alfred Wilkes.[74] The director of the Science Service of the Canadian Department of Agriculture, Kenneth Neatby, pointed out to the Deputy Minister for Agriculture and former professor of animal husbandry at MacDonald College, George Barton, that this orientation was out of place in the federal research laboratories, which should be concentrating on the more advanced and fundamental questions of their sciences.[75] For example, in 1939, when Allison Pickett was appointed director of the laboratory in Annapolis Royal, Nova Scotia, he began immediately to de-emphasize the more empirical work on chemical insecticides. He mounted instead a major program to study the ecology of the Oystershell Scale, the Red Mite, the Eye Spotted Bud Moth, and the Codling Moth, in the expectation that this knowledge would aid the organization of more effective spray calendars for the "supervised control" system which he had helped establish around Nova Scotia during his previous ten years as Provincial Entomologist.[76] The situation in Belleville was remedied by replacing the older generation of entomologists with a newer one, like Pickett, attuned to the changing orientation of entomological work in the Canadian Department of Agriculture.

BIOLOGICAL CONTROL AND ECOLOGICAL RESEARCH

In 1955, Bryan Beirne, a graduate of Trinity College, Dublin, was transferred from the Insect Systematics Unit to become the director of the Biological Control Investigations Unit. He first established a clearer division of responsibilities between the Unit and the Commonwealth Institute for Biological Control, and then proceeded to encourage the interest of new recruits in insect bionomics and ecology rather than the business of importing and distributing parasites and predators from abroad.[77] Some of them were sent abroad, with the financial assistance of the Department of Agriculture and the National Research Council, for doctoral training in these different aspects of entomology. Thus, Albert Turnbull went to Oxford University to study the ecology of spiders under the leading population ecologist George Varley, and his colleague Donald Chant obtained his doctoral degree under Arthur Massee at the University of London for work on the systematics and bionomics of Spider Mites. This changing orientation of the Biological Control Investigations Unit was given the official stamp of approval as the latter was upgraded in 1959 to become the Entomology Research Institute for Biological Control.[78]

The Biological Control Investigations Unit was not the only center for research in biological control to undergo this transformation. The Forest Insects Investigation Unit in Fredericton, which was administered by the Department of Agriculture until 1959, when it was transferred to the newly established Department of Forestry, had worked quite extensively on the development of biological control programs to protect Canadian lumber.[79] Controlling insects attacking forests seemed so important that, by the 1950s, even half the staff at Belleville worked in this area.[80] All this work, especially that which focused on the ecology of insect populations, was strongly supported by the lumber and wood pulp industries well into the 1950s, when Fraser Companies and the Canadian Pulp and Paper Association underwrote the "Green River Project" to study the population dynamics of the Spruce Budworm. This project was fundamentally important for a group of Canadian population ecologists, whose work was centrally important to debates over the theoretical foundations of biological control, and ecology generally.[81] Thus, when Crawford "Buzz" Holling, from the Department of Forestry, received in 1966 the prestigious Mercer Award, which was granted annually by the Ecological Society of America for the best ecological research publication of the year, the award committee suggested that Holling's work on the ecology of predation had been made possible by his employers' "enlightened research policy".[82]

Entomologists in the Biological Control Investigations Unit and in the Forest Insects Investigation Unit, like scientists in all the other federal research centers across Canada, were being encouraged, in brief, to explore the scientific foundations of particular technological practices rather than any operational problems. As Beirne told his audience at a meeting of the Entomological Society of Ontario in 1962, they were being handed great latitude to develop their research program as they themselves thought might best advance the more practical operations for which they were nominally responsible. The relationship between these programs and the work of entomologists involved with the more operational phases of crop protection however, still received very little attention, as it fell outside the formal brief of research institutes such as that in Belleville. It is perhaps not surprising, then, that for all their technological orientation the new ideas generated by the entomologists in the Entomology Research Institute for Biological Control found little application in the field. Those responsible for operational phases, Beirne explained, were not well informed about them.[83] However, considering the history of Pickett's pioneering work in integrated control, we may speculate that their ideas may have seemed, as Beirne put it, "abstruse" and "esoteric" to the more practical economic entomologists because they were conceived without any real sensitivity for the constraints imposed by agricultural practice.

DEVELOPING AN INTEGRATED CONTROL APPROACH

Robert White Stevens' vitriolic criticism of *Silent Spring* in the *Agricultural Institute Review* can leave no doubt that the attitude of the insecticide industry toward Canadian work in biological control was just as critical as that of that in the United States.[84] In the federal laboratories, however, there was only limited competition for funding between entomologists interested in biological and chemical control. The principal center for research in the latter area, the Pesticide Research Institute in London, Ontario, was established in 1951 to investigate the problems created by the introduction of synthetic insecticides like DDT, and, just as in the Entomology Research Institute for Biological Control, research there focused more on the physiological mechanisms of insecticidal action and the genetics of resistance, than on the practical development of insecticidal programs.[85] This institute certainly was larger than that in Belleville, but this was a far better situation than in the Division of Entomology in the United States Department of Agriculture, where work in biological control had virtually come to a standstill after the second world war. With this more even-handed approach to biological and chemical control, it was relatively easier to study and develop programs to integrate the two.[86] Thus, even as early as in 1947, when the Entomological Society of Ontario held a symposium on "Problems of Orchard Pest Control", a number of speakers, and Allison Pickett in particular, recommended the development of such integrated programs, if only to guarantee that DDT would continue to be an effective insecticide longer than arsenical ones had been.[87]

During the economic depression in the 1930s, apple growers in Nova Scotia, just like citrus growers in southern California, were forced to lower their costs of production to stay in business. However, unlike citrus growers, these apple growers increased their use of insecticides dramatically to eliminate all losses to the Apple Maggot and the Codling Moth. What they found as they did so, was that the cost of protection continued to climb and their losses appeared only to worsen. By the early 1940s they had ceased to use any insecticides at all, and had virtually given up on the business of growing apples. Pickett and his associates at the Annapolis Royal became convinced that there was an interesting problem in Nova Scotian apple orchards which deserved closer investigation. Between 1946 and 1953, they produced a number of papers on the ecology of insects in apple orchards.[88] Some of the entomologists in Pickett's group concerned themselves strictly with the toxic effects of different insecticides on predators and parasites controlling insect species attacking apple crops, and eventually found that some insecticides, such as Ryania, were far less toxic to some species than to other ones.[89] Pickett himself, along with his colleague Nelson Patterson, investigated how all these different insecticides affected the population dynamics of the insect species attacking the crop and of their natural enemies.[90] They discovered that each insecticide had different and specific effects on these ecological dynamics, as did the use of the same insecticide during different phases of the interactions between insects and their natural enemies. They also found that the use of insecticides could sometimes encourage the growth of those insect populations which they were meant to eliminate. Based on all this work, Pickett and his team suggested that the use of pesticides, when informed by a detailed knowledge of insect population dynamics, could result in better control and reduced need for them. They warned, however, that long-term ecological investigations to obtain this knowledge would have to be undertaken to adopt the approach, and that opportunities for them would be few. They wrote, in the *Canadian Entomologist*, that,

> comprehensive studies on the ecological relationships of the species involved and on the immediate and long-term consequences of artificial controls are the only alternatives to costly interference with the forces of nature ... The empirical methods that economic entomologists have been obliged to use are frequently of dubious value and should be considered as temporary measures that must be continuously reexamined. It is easy to say that this or that should or should not have been done, but the fact remains that at no place in the world have the necessary facilities been available to carry on the very extensive and intensive research programs necessary to give proper insight into the phenomena with which we are dealing. Our own feeble attempts are but a drop in the bucket.[91]

For the entomologists at Annapolis Royal, the lack of ecological insight was the major hurdle in the way of a wider adoption of an integrated approach to crop protections. They ignored, however, that the program they were proposing rested on the repeated use of Ryania, which was far more expensive that DDT. Moreover, since the Spider Mites attacking the apple crop were not affected by Ryania, it had to be used in conjunction with other insecticides and the costs grew still more.[92] Pickett and his colleagues responded to this difficult situation simply by arguing that growers should accept some limited amount of damage to their apples if they wanted to avoid the supposedly more costly disruptions caused by DDT.[93] Such a view would have been untenable for any entomologists more closely associated with farmers and growers, such as those in the Citrus Experiment Station at Riverside.[94] Not surprisingly, Pickett's work was not followed by widespread adoption of integrated control.[95] It followed the path of biological control, which was neglected and not widely applied anywhere in Canada. Chemical insecticides continued to be preferred by growers, and, rather unsurprisingly, the impact of DDT was as disastrous in Canada as it was in the United States.

REORGANIZING RESEARCH IN BIOLOGICAL CONTROL

Failures such as those of the groups at Annapolis Royal and Belleville to transform Canadian farmers' practices of crop protection, along with those of other Canadian agricultural research institutes generally to fulfill their diverse agricultural missions, motivated, as we saw in the last chapter, numerous discussions during the late 1950s about the need to re-organize these research institutes. Administrators in the Department of Agriculture, such as Kenneth Neatby and Cyril Goulden, believed that the organization of these institutes along lines of disciplinary expertise, and their separation from the Experimental Farm Service, were the major factors hampering their formal mission.[96] The director of the Biological Control Investigations Unit, Bryan Beirne, was aware of similar problems in Belleville. If farmers' seemingly self-defeating reliance on chemical insecticides was to be stopped, he believed, that a much broader approach to the problems of crop protection was needed. Entomologists such as Donald Chant were encouraged, therefore, to develop their interest in integrated control. However, they were unable fully to pursue this line because work on insecticides and biological control were segregated in different research institutes — this was just the sort of situation that had motivated the institutional reorganization of agricultural research in 1959.[97] The *Research Reports* for 1963 and 1966 readily reveal that, although research in the newly renamed Entomology Research Institute for Biological Control included studies of cultural methods of controlling insect populations, little was done on the integration of this expanded approach to biological control and the use of chemical insecticides. Even when they moved into situations which brought them in closer contact with entomologists more interested in the operational phases of chemical control, entomologists from Belleville continued to maintain a more removed approach to their new responsibilities. For example, in 1960, Chant left Belleville to become the director of the Research Laboratory in Vineland, Ontario. There he worried about the increasing pressure to take up more economically relevant work, and became more directly involved with problems of integrated control, but, for the most part, he continued to focus his attention on the systematics of Spider Mites and the theoretical foundations of biological control.[98]

The tradition of disciplinary research fostered by the universities in which many entomologists in the Department of Agriculture were trained, were deeply entrenched and difficult to overcome. Moreover, those such as Beirne, who were committed to breaking this tradition, to promote more multi-disciplinary work, lacked the close links to university departments needed to transform the educational curriculum of the newer generations of entomologists emerging from Canadian universities. They sought to improve their situation during the 1960s, by trying to build closer links with these universities, but their efforts faltered as the government began to reassign the responsibility for the agricultural research from the Research Branch of the Department of Agriculture to the universities.[99] The number of federal entomologists who then left for posts in the United States, or took advantage of the improving opportunities for research in Canadian universities, was so great that Beirne found it very difficult to staff and continue work in Belleville. Finally, in 1967, he took advantage of new federal "development grants" to promote research in the universities, and moved with eight of his personnel to the recently established Simon Fraser University, to establish there the first Canadian graduate course in pest management.[100] Philip Corbett, an insect ecologist trained at Cambridge University, replaced Beirne, but soon resigned to take up a post at Guelph University.

Following Corbett's resignation, the status of the Entomology Research Institute for Biological Control was reviewed by administrators in the Department of Agriculture. One of the members of the reviewing panel, the insect ecologist and systematist, Eugene Munroe, strongly criticized the work done at Belleville. Taking as an example a critique of practices in biological control by Chant and Albert Turnbull, a critique which I will soon consider in some detail, Munroe argued that it inclined too much toward study of strictly entomological problems, rather than broader agro-ecological ones, and was out of touch with practical developments.[101] He felt that this approach was more appropriate for scientists in universities, than in the "mission-oriented" laboratories of the Department of Agriculture.[102] Canadian work in biological control should be structured as it was in Riverside and Berkeley, where, Munroe believed, there was an exemplary integration of research, development and operation.[103] The outcome of this was the disbanding of the Institute, and the dispersal of its remaining staff was dispersed in various laboratories across Canada.

In sum, the histories of institutional arrangements for work in biological control in the United States and Canada were radically different. Under the leadership of Harry Smith, entomologists in the Division of Biological Control in the Citrus Experiment Station at Riverside always maintained close and strong ties with their institutional patron, the citrus industry of southern California. Gordon Hewitt and his successors, instead, succeeded in building a Biological Control Investigations Unit at Belleville which was relatively independent from its nominal patrons, agricultural producers throughout the country. Their most important patron was the Canadian Department of Agriculture in Ottawa. Agricultural scientists commanded important positions in this Department, positions from which they could shape the institutional development of agricultural research as they thought it best. These relationships continued virtually unchanged into the 1950s. Not surprisingly, during these years entomologists in Riverside, Berkeley and Belleville viewed the relationship between their scientific work and the operational development of control programs quite differently. This, as we shall see in the next chapter, had very specific implications for their ideas about the ecological foundations of work in biological and integrated control.

NOTES

1 R. Sawyer, "To make a spotless orange: biological control in California", (Ph.D. dissertation, University of Wisconsin, 1990), pp. 21–4; see also "Monopolizing the insect trade: biological control in the USDA, 1881–1951", *Agricultural History* 64 (1990): 271–85.

2 L.O. Howard, *A History of Applied Entomology* (Washington: Smithsonian Institution, 1930), p. 51.

3 Howard, *History*, p. 500.

4 Sawyer, "To make a spotless orange", p. 25; see also R. Doutt, "Vice, virtue, and the Vedalia", *Bulletin of the Entomological Society of America* 4 (1958): 119–123.

5 L.O. Howard & W.F. Fiske, *The Importation into the United States of the Parasites of the Gypsy Moth and the Brown Tail Moth*, United States Department of Agriculture Bureau of Entomology Bulletin No. 91 (1911), pp. 30–46.

6 See R.P. McIntosh, *The Background to Ecology. Concept and Theory* (Cambridge: Cambridge University Press, 1985), pp. 160–1 and 189.

7 R.I. Sailer, "A look at United States Department of Agriculture's biological control of insect pests: 1888 to present", *Agricultural Science Review* 10 (1972): 15–27, on pp. 16 and 20.
8 A.R. Grammer, "A history of the experiment station of the Hawaiian Sugar Planters' Association, 1895–1945", *Hawaiian Planters' Record* 51 (1946): 177–228, on p. 193; and C.E. Pemberton, "History of the entomology department (experiment station HSPA) 1904–1945", *Hawaiian Planters' Record* 52 (1948): 53–90, on p. 60.
9 Sawyer, "To make a spotless orange", pp. 33–34; and Pemberton, "History of the entomology department".
10 Sawyer, "To make a spotless orange", pp. 30–34.
11 Sawyer, "To make a spotless orange", p. 7.
12 Sawyer, "To make a spotless orange", pp. 96–100.
13 Sawyer, "To make a spotless orange", pp. 79.
14 Sawyer, "To make a spotless orange", p. 115; on agricultural scientists in the University of California, see B.A. Kimmelman, "A Progressive era discipline: genetics and American agricultural colleges and experiment stations, 1890–1920", (Ph.D. dissertation, University of Pennsylvania, 1987), pp. 295–302.
15 Sawyer, "To make a spotless orange", pp. 112–5.
16 Sawyer, "To make a spotless orange", p. 79.
17 Sawyer, "To make a spotless orange", pp. 81–7 and 104.
18 Sawyer, "To make a spotless orange", pp. 116–7.
19 Sawyer, "To make a spotless orange", pp. 124 and 157.
20 Sawyer, "To make a spotless orange", p. 138.
21 Sawyer, "To make a spotless orange", pp. 124–5.
22 R.K. Vickery, "Pest control and the New Deal", *Journal of Economic Entomology* 28 (1935): 1078–9.
23 Sawyer, "To make a spotless orange", pp. 141 and 146.
24 P. De Bach & H.S. Smith, "Are population oscillations inherent in nature?" *Ecology* 22 (1941): 363–9.
25 Sawyer, "To make a spotless orange", pp. 137 and 179.
26 H.S. Smith, "Racial segregation in insect populations and its significance in applied entomology", *Journal of Economic Entomology*, 34 (1941): 1–13.
27 Sawyer, "To make a spotless orange", pp. 183–4.
28 Sawyer, "To make a spotless orange", p. 226; see also P.N. Annand, "Preventive entomology", *Journal of Economic Entomology* 40 (1947): 461–8.
29 A.E. Michelbacher, "The importance of ecology in insect control", *Journal of Economic Entomology* 38 (1945): 129–30.
30 See R. Woglum *et al.*, "The effect of field-applied insecticides on beneficial insects of citrus in California", *Journal of Economic Entomology* 40 (1947): 818–20.

31 Sawyer, "To make a spotless orange", pp. 201–6.
32 Sawyer, "To make a spotless orange", pp. 217–8.
33 Sawyer, "To make a spotless orange", p. 224.
34 Ken Hagen (University of California), interview with Paolo Palladino, 6 July 1988; and Sawyer, "To make a spotless orange", pp. 205–6 and 226.
35 T.R. Dunlap, *DDT: Scientists, Citizens, and Public Policy* (Princeton: Princeton University Press, 1981), p. 64.
36 V.M. Stern & R. Van den Bosch, "Field experiments on the effects of insecticides", *Hilgardia* 29 (1959): 103–30.
37 Sawyer, "To make a spotless orange", pp. 232–3.
38 V.M. Stern *et al.*, "The integrated control concept", *Hilgardia* 29 (1959): 81–101.
39 B.R. Bartlett, "Laboratory studies on selective aphicides favoring natural enemies of the spotted alfalfa aphid", *Journal of Economic Entomology* 51 (1953): 374–8.
40 C.B. Huffaker, "Ecological management of pest systems" in J.A. Behnke (ed.), *Challenging Biological Problems: Directions towards their Solution* (New York: Oxford University Press, 1972), pp. 313–42, on p. 319.
41 Stern, "Integrated control", p. 97.
42 Sawyer, "To make a spotless orange", pp. 237–44. See also N.A. Kerr, *The Legacy: A Centennial History of the State Agricultural Experiment Stations, 1887–1987* (Missouri Agricultural Experiment Station, 1987), pp. 94–101.
43 Sawyer, "To make a spotless orange", pp. 209–12.
44 See C.B. Huffaker, "Experimental studies on predation: disperson factors and predator-prey oscillations", *Hilgardia* 27 (1958): 343–83; and P. De Bach, "The competitive displacement and coexistence principles", *Annual Review of Entomology* 11 (1966): 183–212.
45 Sawyer, "To make a spotless orange", p. 238.
46 See M. Kogan, "Preface", in Kogan (ed.), *Ecological Theory and Integrated Pest Management Practice* (New York: Wiley, 1986), p. ix.
47 R. Glen, "Entomology in Canada up to 1956: a review of developments and accomplishments", *Canadian Entomologist* 88 (1956): 290–371, on p. 364.
48 B.P. Beirne, "Influences on the development and evolution of biological control in Canada", *Bulletin of the Entomological Society of Canada* 5 (1973): 85–9.
49 A.B. Baird, "A brief survey of the organization and projects of the Belleville parasite laboratory", *Annual Report of the Entomological Society of Ontario* 61 (1930): 65–8; and P.W. Riegert, *From Arsenic to DDT: A History of Entomology in Western Canada* (Toronto: Toronto University Press, 1980), p. 88.
50 See C.G. Hewitt, "Some observations on the practical importance of the study of parasitic insects", *Proceedings of the Entomological Society of Ontario* 41 (1910): 62–4.

51 A.B. Baird, "The Entomological Society and biological control in Canada", *Proceedings of the Entomological Society of Ontario* 77 (1946): 5–7.
52 L.O. Howard, "History", p. 415; and Glen, "Entomology in Canada", p. 318.
53 Riegert, *Arsenic to DDT*, pp. 88 and 115.
54 Beirne, "Influences", p. 85.
55 National Archives of Canada (Ottawa): record group 17: volume 3049: folder 40–17(1): A. Gibson to J.H. Grisdale, 24 January 1922.
56 A. Gibson, "The development of applied entomology in Canada, 1914–1923", *Report of the Quebec Society for the Protection of Plants* (1924): 24–56; see also National Archives of Canada: rec. group 17: vol. 3049: fold. 40–17(1): A. Gibson to J.H. Grisdale, 24 January 1922.
57 Baird, "Entomological Society and biological control"; and Beirne, "Influences", p. 86.
58 W. Downes, "Fifty years of entomology on Vancouver Island", *Proceedings of the Entomological Society of British Columbia* 48 (1951): 9–16.
59 See E.P. Venables, "Fifty years of entomology in the interior of British Columbia", and J. Marshall, "Applied entomology in the orchards of British Columbia", both published in the *Proceedings of the Entomological Society of British Columbia* for 1951, on pp. 19–22 and 25–31. On the growing use of insecticides in Canada more generally, see Gibson, "Applied entomology", pp. 37–8.
60 A.B. Baird, "The biological control of insect pests in Canada", *Proceedings of the Pacific Science Congress* 5 (1933): 3537–42, and Smith on p. 3583.
61 Baird, "Brief survey", p. 66.
62 Beirne, "Influences", p. 86.
63 National Archives of Canada: rec. group 17: vol. 3038: fold. 40–1(8): Proceedings of "Conference on biological methods of controlling pests", 25–26 June 1936, pp. 67–69.
64 See J.P. Hull, "From the FPL to PAPRICAN: science and the pulp and paper industry", *HSTC Bulletin* 8 (1983): 3–13.
65 Library of the Canadian Department of Agriculture (Ottawa): A. Gibson, *Entomology in Canada, with Special reference to Developments in the Dominion Department of Agriculture*, unpublished manuscript, 1951, p. 99.
66 A.B. Baird, "Biological control of insect pests in Canada with specific reference to the control of the European Spruce Sawfly", *Proceedings of the Entomological Society of Ontario* 70 (1939): 51–56, on p. 54; see also Glen, "Entomology in Canada", pp. 352–353.
67 National Archives of Canada: rec. group 17: vol. 3047: fold. 40–6(1): R.A. McInnis to J.G. Gardiner, 2 February 1937, and Minutes of meeting to consider the European Spruce Sawfly situation, 29 October 1937.
68 Gibson, *Entomology in Canada*, p. 143.
69 L.S. McLaine, "The war activities of the federal Division of Entomology and Plant Protection", *Proceedings of the Entomological Society of Ontario* 73 (1942): 7–16.

70 National Archives of Canada: rec. group 17: vol. 3047: fold. 40–6(2); and vol. 3046: fold. 40–5–70.
71 Baird, "Biological control", pp. 3539–42.
72 Bryan Beirne (Simon Fraser University) and Ken Watt (University of California), interviews with Paolo Palladino, 12 June 1992 and 24 July 1991 respectively. On Park's ideas about the ecology of animal populations, see G.A. Mitman, *The State of Nature: Ecology, Community, and American Social Thought, 1900–1950* (Chicago: University of Chicago Press, 1992).
73 Bryan Beirne (Simon Fraser University), interview with Paolo Palladino, 7 July 1988; and Beirne to Paolo Palladino, 13 May 1992. See also Thompson in R.E. Balch, "Symposium on entomological education", *Annual Report of the Entomological Society of Ontario* 79 (1948): 58–65, on p. 63.
74 Canada Department of Agriculture, *Research Reports: Research Institute, Belleville, Ontario (1956–1959)* (Ottawa: Queen's Printer, 1960), p. 9.
75 National Archives of Canada: rec. group 17: vol. 3045: fold. 40–5–49–1: K.W. Neatby to H. Barton, 1 October 1947.
76 Glen, "Entomology in Canada", p. 335; and J.F. Hockey, *Agricultural Research in the Annapolis Valley Area, 1909–1965* (Ottawa: Queen's Printer 1967), p. 35.
77 Bryan Beirne to Paolo Palladino, 13 May 1992.
78 Beirne, "Influences", pp. 86–7.
79 Glen, "Entomology in Canada", p. 355.
80 Beirne, "Influences", p. 87.
81 Glen, "Entomology in Canada", pp. 317–22, 355 and 361; and G.J. Spencer, "A century of entomology in Canada", *Canadian Entomologist* 96 (1964): 33–59, on pp. 53–5.
82 "George Mercer award for 1966", *Bulletin of the Ecological Society of America* 47 (1966): 168–9, on p. 168.
83 B.P. Beirne, "How useful is the basic research program to economic entomology? Entomology Research Institute for Biological Control, Belleville", *Proceedings of the Entomological Society of Ontario* 93 (1962): 5–6; and "Biological control of insects", *Pesticide Progress* 1 (1963): 24–7.
84 R. White Stevens, "Feeding an exploding population: the role of agricultural chemicals", *Agricultural Institute Review* 17 (1962): 11–3 and 46.
85 See H. Martin & R.D. O'Brien, "The Science Service Laboratory, London (Canada)", *Laboratory Practice* 6 (1957): 339–46; and Canada Department of Agriculture, *Research Reports: Pesticide Research Institute, London, Ontario (1957–1960)* (Ottawa: Queen's Printer, 1961), p. 9.
86 Donald Chant (Ontario Waste Management Corporation), interview wiith Paolo Palladino, 17 August 1988.
87 See J. Marshall, "Pest control in British Columbia orchards", H.R. Boyce, "The role of biological control in Canadian orchard entomology", W.L. Putnam, "Problems in orchard insect control as exemplified by the

Oriental Fruit Moth", and A.D. Pickett, "The philosophy of orchard insect control", all published in the *Proceedings of the Entomological Society of Ontario* for 1948, on pp. 26–41.

88 F.T. Lord, "The development of agricultural entomology in Nova Scotia", *Proceedings of the Nova Scotia Institute of Science* 33 (1986): 19–46, pp. 40–2.

89 Lord, "Development of agricultural entomology", p. 43.

90 A.D. Pickett & N.A. Patterson, "The influence of spray programs on the fauna of apple orchards in Nova Scotia. IV. A review", *Canadian Entomologist* 85 (1953): 472–8; see also D.A. Chant, "Some highlights at the tenth International Congress of Entomology: biological control", *Proceedings of the Entomological Society of Ontario* 87 (1956): 54–8.

91 Pickett & Patterson, "Influence", pp. 472–3.

92 G.G. Dunstan, "Some highlights at the tenth International Congress of Entomology: fruit insects", *Proceedings of the Entomological Society of Ontario* 87 (1956): 58–61.

93 Lord, "Development of agricultural entomology", p. 44; and T.H. Anstey, *One Hundred Harvests: Research Branch, Agriculture Canada, 1886–1986* (Ottawa: Research Branch, Agriculture Canada, 1986), p. 271.

94 Sawyer, "To make a spotless orange", pp. 234–5.

95 G.G. Dunstan & R.S. Willison, *A History of the Entomology and Plant Pathology Laboratory on the Niagara Peninsula, 1911–1960* (Ottawa: Queen's Printer, 1968), pp. 18–9.

96 Anstey, *One Hundred Harvests*, p. 90.

97 Chant, "Highlights"; and Beirne, "Basic research program".

98 Canada Department of Agriculture, *Research Reports: Research Laboratory, Vineland, Ontario (1958–1962)* (Ottawa: Queen's Printer, 1963), p. 6.

99 Bryan Beirne to Paolo Palladino, 18 May 1992; see also D.K.M. Kevan, "The department of entomology at McGill University: a history to 1978", *Notes of the Lyman Entomological Museum Research Laboratory* 5 (1979): 1–88, on p. 30.

100 P.A. Larkin & W.J.D. Stephen, *From Formalin to Fortran* (Ottawa: Science Council of Canada, 1971), p. 42.

101 E.G. Munroe, "Status and potential of biological control in Canada" in "Biological control programmes against insects and weeds in Canada, 1959–1968", *Commonwealth Institute of Biological Control, Technical Communications* 4 (1971): 213–55, on pp. 235–6, 243–6 and 252; see also A.L. Turnbull & D.A. Chant, "The practice and theory of biological control of insects in Canada," *Canadian Journal of Zoology* 39 (1961): 697–751.

102 Munroe, "Biological control in Canada", pp. 216 and 248.

103 Munroe, "Biological control in Canada", p. 251.

CHAPTER 4

Biological Control, Integrated Control, and Ecological Theory

CHEMICAL, BIOLOGICAL, AND INTEGRATED CONTROL

In the first two chapters, I sought to explain how the great emphasis placed on the practical utility of all work in economic entomology led American entomology to grow, over the first half of this century, into a virtual sub-field of chemistry, and how its Canadian counterpart followed instead a path to disciplinary establishment more typical of the modern life sciences. Then, in the last chapter, I discussed how the different factors and contingencies that led to these developments were reflected in the changing institutional organization of biological control in the United States and Canada. The most active American center for work in this area, the Division of Biological Control in the Citrus Experiment Station at Riverside (which was later expanded to include a branch at Berkeley), was strongly oriented toward the practical development and implementation of biological control programs. By the late 1950s, however, its ties to agriculture became rather looser, especially as it began to employ a younger generation of entomologists more interested in population ecology and the theory of biological control. Yet, these ties never became as loose as those between the federal Entomology Research Institute for Biological Control at Belleville and Canadian agricultural industries. Work in the latter center focused much more definitely on the ecological foundations of biological control than on any more strictly economic problems. In this fourth chapter, I will consider in some detail the relationships between this institutional difference and the ideas of entomologists in Riverside, Berkeley, and Belleville about how to study the ecology of insect populations, and about the significance of the emergent ecological insight for practices of biological and integrated control.

As we have already seen, during the 1950s, entomologists in these American and Canadian centers sought to develop programs for crop protection which would allow the combined use of chemical and biological control agents. If the former were used in a way that minimized any disruption of the native and introduced biological factors controlling the growth of pest populations, they argued, such an approach would eventually reduce the need for chemical insecticides. In this way, the disastrous consequences of the contemporary, ecologically indiscriminate, and economically senseless use of DDT and other powerful insecticides might become just memories of a distant past. However, careful comparison of the different definitions of the integrated strategy articulated during the early 1960s by entomologists in Berkeley and Riverside, on the one hand, and those in Belleville, on the other, reveals that there was very little agreement among them over the exact importance of biological control in the new strategy. While the former group of entomologists attributed to it great importance, the latter (and others closely associated with them) emphasized the need for a number of supplementary tactics, beside the use of natural enemies and insecticides, to develop any effective integrated control program. The rhetoric of this disagreement, a disagreement which involved, at least in part, a central and endlessly controversial theme of ecological discourse, namely that about the abundance of biological populations and their regulation, suggests that these entomologists' opposing positions were deeply shaped, if not determined, by their different institutional locations.

THEORIES ABOUT THE BALANCE OF NATURE

In late nineteenth century North America, entomologists contributed to well-established discourse about the regulation of natural populations by explaining that insects existed in a tightly and precisely woven web of interactions, like predation and parasitism, and that these interactions acted to maintain the parts of this web in balance with one another. This was usually referred to as the "balance of nature". Human activities, however, disrupted this balance, and hence the reason for the observed depredations of the farmers' crops. These entomologists and their successors, and especially those of them interested in biological control, then went on to elaborate, in many different and more complicated ways, this seemingly simple and unproblematic idea. Chief among were the entomologists John Nicholson and Robin Thompson.[1]

Throughout the 1920s and '30s, Nicholson and Thompson sought to ground the practice of biological control in mathematically rigorous theories of population ecology. Since this practice was based on the use of predatory and parasitic species to control insect pest and weeds, and since not all predatory and parasitic species behaved in the same way, they argued that improving the effectiveness of biological control required a more precise understanding of how these natural enemies affected the abundance of the species upon which they preyed. Thus, when he first joined the Entomology Division of the Australian Council for Scientific and Industrial Research, after having studied zoology at the University of Birmingham and taught at the University of Sydney, Nicholson became very interested in the progress of one of the most successful programs in biological control of weeds ever undertaken. By the beginning of this century, large portions of Australian pastures had been rendered useless by the spread of the prickly pear cactus, which had been introduced there quite accidentally during the late eighteenth century. During the 1920s, entomologists employed by the Commonwealth Prickly Pear Board introduced a moth, *Cactoblastis cactorum*, which soon brought the spread of the weed under control.[2] The dynamics of the moth and cactus populations seemed to confirm ideas about the regulation of natural populations which Nicholson had been developing during these same years.[3]

In 1935, Nicholson, then director of the Division of Economic Entomology of the Council for Scientific and Industrial Research, and Victor Bailey, a physicist and Nicholson's former colleague at the University of Sydney, reformulated and presented in a very abstract, but logically compelling, way the arguments for the existence of a balance of nature already articulated in a more qualitative and approximate way by William Fiske, in that most fertile training center for workers in biological control, the Gypsy Moth Laboratory of the United States Department of Agriculture in Melrose.[4] Nicholson and Bailey suggested that if the reproductive rate of any biological population were less than zero, that is to say, each death were not matched by at least one birth, it would die out sooner or later. If the rate were greater than zero, the population would grow eventually to over-exploit its resources and again die out. The growth rate had to be, therefore, exactly zero for the population to persist over any length of time. What, Nicholson and Bailey then asked, was the mechanism that insured the preservation of such a finely tuned balance between reproductive and mortality rates? According to them, only those factors that slowed population growth with greater or lesser intensity, in strict proportion to the density of the same population, could possibly maintain this balance. Only competition for limited resources, predation, and parasitism, they believed, could possibly act in such a "density dependent" manner. All the other factors affecting the size of a population were "density independent", that is, their limiting effect on population growth was independent of the population size. These factors, Nicholson and Bailey believed, could not possibly maintain such a precarious, but finely tuned balance. Nicholson asked, for example, how climatic factors could act to kill only one individual on the day when only one was born, and act to kill a thousand on that day when one thousand individuals were born. Nicholson and Bailey concluded that density dependent factors "regulated" the population by maintaining its density in equilibrium with its resources, preventing both extinction and indefinite growth. The absolute value of this density, however, was set by environmental constraints; these constraints "determined" the observed abundance of animal populations. In sum, Nicholson argued that physical factors could play no role in shaping the dynamics of animal populations.

Thompson, a former student at Ontario Agricultural College, Cornell University, and the University of Paris, and then embattled director of the Parasite Laboratory at Farnham Royal, would not accept this conclusion. In 1938, he published a blistering criticism of Nicholson's account of the balance of nature.[5] He believed that, if there existed a balancing force that accounted for the persistence of animal populations, this was nothing else but the sum of all the environmental factors impinging on that population. But, given a definition that included everything, and excluded nothing, he failed to see the utility of even postulating the existence of such a balancing force. This view was bolstered by his earlier work on the population dynamics of the Corn Borer while still employed by the United States Department of Agriculture in the European Parasite Laboratory at Hyères. Thompson observed that a mild winter would often result in an explosive growth of the Borer population the following spring. Many times this was not accompanied by an equal growth of the natural enemy population, as would have been expected if Nicholson's analysis had been correct. The growth of the Borer and its natural enemies was affected quite differently by the same environmental circumstances. Hence it seemed to Thompson that population density should be viewed as a function of both favorable climatic conditions and the abundance of natural enemies. On the basis of these considerations, he charged that the distinction between regulating and determining factors could only be the fruit of unbridled mathematical imagination. He was convinced that,

> mathematical theories, though a natural and useful development, have not the significance attributed to them; that none of the theories so far produced can be considered as an adequate representation of events in nature, while the theory at present dominant, being of a form derived from physical phenomena, is particularly inapplicable to the behavior and interaction of living organisms; and that observation and experiment are still the only safe basis for practical operations in economic biology.[6]

The conflicting ideas about the relationship between prey or hosts and their predators or parasites, which Nicholson and Thompson articulated so forcefully, could have important consequences for the establishment of more effective biological control because entomologists might set higher priority on different characteristics of the ideal natural enemy, depending on which ecological perspective they adopted. Those who tended to share Nicholson's views, would also tend to believe that the more effective natural enemies were those that searched for hosts or prey over a greater area, because natural enemies would have greater chances of finding their prey or host at very low densities and thus drive the latter species' abundance to very low levels. On the other hand, those entomologists who shared Thompson's views might tend to search for those natural enemies whose reproductive rate, relative to that of the prey or host, was greatest. Since they would have believed, following Thompson, that the population density of the natural enemies was at least partly fixed by the vagaries of climate, it was important that the natural enemy should be capable of reproducing very quickly to match to any outbreaks of its prey caused by favorable climatic spells. More fundamentally, however, since the determination of insect numbers was only partially attributable to the role of natural enemies, this latter group of entomologists also tended to be rather skeptical of the optimistic claims about the effectiveness of biological control emerging from, for example, the Citrus Experiment Station in Riverside, California. To them, and especially to the Chief of the Bureau of Entomology of the United States Department of Agriculture, Leland Howard, the success in California with the Vedalia Beetle was an exception rather than the rule. Thus, they thought that chemical insecticides and cultural modifications, such as the breeding corn resistant to infestations by the Corn Borer encouraged by Thompson, were bound to become the mainstay of work in crop protection.[7]

BIOLOGICAL CONTROL, INTEGRATED CONTROL, AND ECOLOGICAL THEORY IN RIVERSIDE AND BERKELEY

By the time of the debates between John Nicholson and Robin Thompson over ecological theory and biological control, Harry Smith had established in the Citrus Experiment Station at Riverside strong institutional support for biological control. According to him and the members of the school of biological control which was developing around him, insect and weed populations grew to economically significant densities because they were either exotic species, lacking their complement of natural enemies, or because the crop was exotic, thus providing a novel and unexploited niche for native insect and weed species. What workers in biological control had to do, then, was to establish a new, balanced host-parasite or predator-prey complex and growers' problems would promptly disappear. That is to say, Smith viewed their problem as one of re-establishing the balance of nature.[8] Not surprisingly, therefore, Nicholson's ideas on the regulation of animal populations seemed very attractive to the entomologists in the Division of Biological Control in Citrus Experiment Station. In fact, in 1935, Smith suggested; in a paper on the dynamics of insect populations, that Nicholson's theory, with its emphasis on the fundamental role of natural enemies in the regulation of animal populations, fitted well with his own views on the origin of insect problems.[9] Then, in 1942, he and his student, Paul DeBach, published in *Ecology* the first experimental evidence for the theory, something which Nicholson himself had despaired of ever obtaining (and a testament to these entomologists' great skills in experimental work).[10] According to Nicholson's theory, the temporal lag between the growth of the host or prey population and that of its natural enemies should cause the densities of the two populations to fluctuate cyclically about some equilibrium value, which was set by environmental factors. Smith and DeBach thought that this aspect of the theory accounted for the fluctuations they observed in the experimental interaction of a population of common houseflies and its pupal parasite, *Mormoniella vitripennis*. This intellectual connection with Nicholson soon became a personal one as the latter began to pay a number of visits to the Citrus Experiment Station beginning in the 1940s and continuing into the 1950s.

However, for Smith and for many of his colleagues in Riverside, and then Berkeley, the most striking evidence in favor of Nicholson's theory was not to be found in experimental studies, but was instead offered by the disastrous effects of synthetic insecticides on the ecology of both noxious and beneficial insect species. For example, during the late 1940s, the use of DDT to control the Citricola Scale led to a resurgence of the Cottony Cushion Scale in citrus groves throughout the San Joaquin Valley. These entomologists explained that the unexpected and dramatic growth of the last scale was due to the removal of its chief natural enemy, the Vedalia Beetle. In the absence of this restraint, which entomologists had introduced into California at the end of the nineteenth century to control the Cottony Cushion Scale, the latter could freely reproduce until the ultimate restraint, intra-specific competition for limited resources intervened. Unfortunately for the growers, this limited resource was the citrus crop itself. In other situations, insect species like the Cotton Leafworm, which before the mid-1950s had usually been ignored by cotton growers in the Imperial Valley, began to cause serious damage following intensive use of DDT and BHC against the Pink Bollworm. Here it seemed that the Bollworm was a better exploiter of the cotton plant than the Leafworm, and thus it acted indirectly to maintain Leafworm populations at low levels. However, the Bollworm, by virtue of being the specific target of insecticidal control operations, was affected by the insecticides far more adversely than the Leafworm, and Leafworm populations could therefore grow to greater densities than would have been otherwise possible.

Given these striking effects of the predators and parasites introduced to control such destructive species as the Cottony Cushion Scale, the impact of chemical insecticides on the ecology of noxious and beneficial insect species, and the examination of all these phenomena from a legitimating theoretical perspective, the entomologists in Riverside and Berkeley had no doubts about the central role of natural enemies in regulating the density of animal populations generally. This commitment to the concept of a balance of nature played an important role in the definition of the integrated control strategy proposed by Smith's heirs.

In 1959, Vernon Stern, Ken Hagen, Ray Smith, and Robert Van den Bosch co-authored a paper on "The integrated control concept" for the official publication of the state agricultural experiment station for California, *Hilgardia* — in other words, their primary audience was agricultural scientists. They argued that insect or weed species became the target of control programs for any one of three reasons. The species was exotic, and hence lacked its proper complement of natural enemies. Alternatively, the crop was an exotic species that provided new and unoccupied niches for native insect species, and thus allowed the insects to obtain higher densities than would be possible otherwise. Lastly, changed economic circumstances might require lowering the level of tolerable losses, the so-called "economic threshold"; insect or weed species that used to be ignored because they did not cause significant economic damage would then become targets for control programs. Stern and his colleagues suggested that the first two problems could be addressed by establishing complexes of natural enemies that brought the noxious species under control. To confront the last, more difficult case, they suggested that lowering the equilibrium density of the pest population below the level at which economic damage occurs would resolve the problem. Again, this could be accomplished by introducing a new and more effective natural enemy, or by modifying cultural practices that seemed to hinder the action of existing natural enemies. These entomologists then turned their attention to the destruction of natural enemies by the usual indiscriminate use of chemical insecticides, and its problematic economic consequences. They argued that the latter could be minimized by realizing that, more often than not, the abundance of any potential pest was below the economic threshold, and that these pests were responsible for economically significant damage only during the periodic upswings of population growth. Hence the use of chemical insecticides was only required when these upswings approached, and promised to rise above the treatment threshold. In a nutshell, then, biological control was the central component of integrated control, and chemical control played only a marginal role. In fact, according to Hagen, the ultimate, if not explicit aim of the strategy was to obtain a "self-sustaining" program of pest control, that is, a program that did not require the continued intervention of humans to maintain the balance between human needs and pest depredations in favor of the former.[11] This ultimate goal reveals the unlimited and persistent faith these entomologists in Riverside and Berkeley had in the effectiveness of biological control, a faith that led to much opposition in some quarters of the entomological community.

A DIFFERENT PERSPECTIVE ON BIOLOGICAL AND INTEGRATED CONTROL

In 1964, a former member of staff in Entomology Research Institute for Biological Control at Belleville, and the then director of the Division of Biological Control in Riverside, Donald Chant, presented an alternative definition of integrated control that was very critical of the one formulated by Vernon Stern and his colleagues. He wrote that,

> several important papers have been published on integrated control in the last few years ... Two of the most extensive papers were those of Stern *et al.* (1959) and Van den Bosch and Stern (1962) ... Van den Bosch and Stern (1962) admitted that previous definitions of integrated control have been a combination of chemical and biological control. They agreed with Franz (1961) that this should be broadened to include all artificial pest controls and cultural practices as well as biological control. Like him, they then proceed to discuss only classical chemical and biological control.[12]

More than anything else, Chant objected to the view that biological control should be treated as the centrally important component of an integrated control strategy. He thought that, if any one component should be characterized as central, this should be chemical control. This very different view was not motivated by belief that the use of chemical insecticides alone was at all satisfactory; on the contrary, Chant was extremely critical of their current use and an active participant in the movement to ban the use of DDT.[13] He believed, instead, that the modern development of agriculture left little room for simple solutions devised in the late nineteenth century, such the introduction of the Vedalia Beetle against the Cottony Cushion Scale. Thus, he argued that,

> most authors think of chemical control as a supplement to biological control at times and places where it is inadequate. This concept is valid for forests, where a natural situation usually still prevails. But I do not think it is true in agriculture, especially horticulture. The natural situation in orchards, for example, and for the foreseeable future is for pests to be repressed by chemical controls. The times when biological controls were the dominant regulators in these places are lost in most cases in the past, and no matter how much we may wish it, they cannot be resurrected unchanged. That opportunity was lost when the first pesticides were used. Therefore, we must now look to supplementing chemical controls with biological controls, or achieving control and regulation in other ways. We should not think in the usual way of "restoring natural balance", but rather of creating something entirely new that will be an effective substitute for that which we have lost.[14]

In other words, it seemed to Chant that many entomologists interested in biological and integrated control tended both to view farmland and fruit orchards as assemblages of biological species that were essentially identical to natural ecosystems, and to assume that it was still possible to contain the damage caused by pest populations by simply establishing within these agricultural ecosystems the same mechanisms that supposedly checked the growth of insect populations in natural ones. He rejected these views as nonsensical.

This contrary position was grounded in a study which Chant and Albert Turnbull had prepared in the late 1950s, when still colleagues in Belleville, at the behest of the Entomological Society of America for a joint meeting with the Entomological Society of Canada. They were asked to review the history of attempts to obtain the economic control of insects damaging Canadian crops and lumber by introducing species that preyed upon or parasitized these insects.[15] In their report, which, very significantly, was eventually published in the *Canadian Journal of Zoology* rather than in any entomological, let alone, agricultural publication, they concluded that the record was poor. They explained that this was because, firstly, agricultural ecosystems were inherently unstable. Human necessity had led to such a reduction of the number of insect and plant species found in these ecosystems that potential natural enemies lacked the plant refuges required to thrive and control damaging species. The latter were then able to obtain very high population densities thanks to the large and easily accessible food, the crop. Insect outbreaks were not anomalous situations, but were, instead, to be expected in such artificially simplified ecological communities.

Secondly, Turnbull and Chant were troubled by the association of ecological theory and practices of biological control drawn by the leading proponents of biological and integrated control, the entomologists in Riverside and Berkeley. In their view this was a very loose and unsystematic allegiance to John Nicholson's theory. They drew attention to their practice of introducing a number of different species of natural enemies, such as the three species of *Aphytis*, to control the California Red Scale. Harry Smith and his followers, especially Paul DeBach, hoped that either the combined action of these different species would ultimately bring the scale down to economically acceptable densities, or that the superior searcher among them would eventually displace the others and do the job by itself.[16] Turnbull and Chant pointed out, however, that according to Nicholson's theory the competitive interference between multiple species of natural enemies would inevitably reduce the effective control of the host or prey population.[17] Biological control workers, they argued, should seek out the one most effective natural enemy, and refrain from any introductions until this species was discovered. The fact that the entomologists in Riverside and Berkeley persisted in this practice, despite their awareness of its problematic theoretical status, suggested that their practices were founded in theory only very superficially. This, for the two Canadians, was a recipe for failure, since sound practice could only result from the systematic application of theoretical insight. This notion that theory must always guide practice was so engraved in Chant and Turnbull's minds that, in 1967, the latter surmised incorrectly, and much to the frustration of the entomologists in Riverside and Berkeley, that this practice of multiple introductions must have been motivated by a commitment to the then prevalent notion that increasing the complexity of ecological communities would also increase their stability.[18]

Lastly, and most significantly, Chant and Turnbull argued that the efforts by Smith and his heirs were misguided because they focused too much on evolutionary and ecological questions, that were of little importance to the economic questions that should properly concern them. Biological control, they claimed,

> is sufficiently specialized to warrant consideration as a science in its own right, one that may draw heavily on other disciplines but no more. Its parts should be directed inwards to its own problems, not outwards to the problems of other subjects.[19]

Although the history of the parasites introduced to control the Red Scale was very interesting, when they were first introduced, they competed for the same resource, and, therefore, the practice was, from an economic standpoint, quite undesirable. By persisting with this practice and appealing to evolutionary and ecological theory for justification, as DeBach had done, the two Canadian entomologists suggested, the entomologists in Riverside and Berkeley were pushing biological control work from its proper place outside ecology to the fringes of ecology.[20] This general outlook was summarized in their criticism of Smith's views on the best characteristics of biological control agents:

> Smith judged the efficiency of a parasite by the number of niches it is capable of occupying. We judge it by the number of pest hosts it is capable of parasitizing. Which of these is more nearly correct depends entirely on the point of view with which the parasite is judged. From the ecological, evolutionary viewpoint Smith probably comes closer to the truth; his criterion of efficiency is more likely to produce a 'natural balance'. But ... we are not concerned with a 'natural balance'. We are concerned with a balance that we hope to impose to satisfy our requirements. For this purpose we require quite a different flora and fauna than that which nature provides. Thus we do not necessarily want the parasite that is best adapted to the natural environment; instead we want one that will best do a specific job for us, and that is to prevent certain species from increasing beyond a certain level of abundance.[21]

If biological control was, as Chant and Turnbull claimed, so "sufficiently specialized to warrant consideration as a science in its own right", then the appeal to ecological theory to justify a questionable course of action against pests probably was inappropriate. In fact, Chant was so convinced about this fundamental difference that he suggested, albeit quite tentatively, that, "perhaps we should abandon common concepts of communities, ecosystems, natural balance, and so on, and recognize that what we are trying to do is completely unnatural and artificial".[22] More conservatively, in their co-authored paper, Turnbull and Chant claimed that the limited successes of biological control work indicated, at most, that entomologists were sometimes doing something right, but without fully understanding why they occasionally succeeded, and without gaining any useful theoretical insight therefrom. They recommended therefore a more rigorous and careful analysis of the mechanisms determining the abundance of insect populations feeding on agricultural crops. Only such knowledge, they claimed, would permit the development of more effective biological and integrated control programs. In the meantime, integrated control programs should be built around chemical rather than biological control.

This central position of chemical control, however, was not intended to be a permanent one. Chant suggested, in 1964, that the work of a biometrician and an entomologist, respectively in the Statistical Research Service and the Forest Insects Division of the Canadian Department of Agriculture, Ken Watt and Buzz Holling, opened the way toward a totally different approach to crop protection. During the late 1950s and early '60s, Watt and Holling sought to develop new mathematical methods to analyze the dynamics of insect populations in Canadian forests, which would avoid what they felt were the oversimplified views of nature built into the models Nicholson had proposed thirty years earlier. According to Chant, with Holling's and Watt's methods, it would soon be possible to assemble agricultural ecosystems in which the balance between the needs of humans and those of insect species favored the former to a greater degree than was true at the time. But, to accomplish this reorganization of agricultural ecosystems, Turnbull and Chant argued, it was necessary to adopt a much broader view on the problems of crop protection than was common among most entomologists. It was necessary to go beyond the mere study of the interactions of single insect species and their natural enemies, to include the plants upon which they fed and lived, and also the physical environment. Indeed, Chant wrote that, "one must view the pest not as an isolated phenomenon, but as a product of its environment including all major elements, both biotic and a-biotic".[23]

COMPETING IDEAS ABOUT THE THEORY AND PRACTICE OF BIOLOGICAL CONTROL

Albert Turnbull and Donald Chant acknowledged, of course, that it was not necessary for every step in the establishment of biological control to be predicated on a coherent theoretical framework, and they suggested that there was a need for some sort of dialectic between theory and practice. They wrote that,

> biological control workers have two tasks: they must seek the principles that underlie this method of control; and they must at the same time attempt to apply practical control measures. These two tasks impose a degree of dichotomy on our work: there is practical, ad hoc work ... and there are the basic studies of a long-term nature that might be called fundamental ecology. Let us emphasize that we do not consider this dichotomy a bad thing. We feel that the two approaches should be complementary, and we should encourage the interplay between the search for principles and the solution of practical problems ... Practical problems cannot be put aside indefinitely until we have accumulated enough information to attack them on purely rational grounds.[24]

Yet, it seemed to Turnbull and Chant that Harry Smith's heirs had strayed too far toward "ad hoc, or empirical work".[25] Thus they concluded their review of biological control practice by equating the empiricism of the Berkeley and Riverside entomologists with a "casual" approach. They cautioned against the dangers inherent in such an approach by referring to the warning by the British ecologist Charles Elton that, "mass destruction (by chemicals) and the casual releasing of predators and parasites may some day be looked back upon as we do upon the mistakes of the industrial age, the excesses of colonial exploitation or the indiscriminate felling of climax forests".[26]

The entomologists in Riverside and Berkeley were dismayed by such criticism. They had never denied that they sometimes forsook extensive investigation of their problems in order to produce the immediate solutions demanded by their institutional constituency. For example, while discussing the source of the tremendous problems confronting entomologists following the indiscriminate use of synthetic chemical insecticides in the 1940s and '50s, Vernon Stern and his associates wrote in their first definition of integrated control that,

> some of these problems have arisen from our limited knowledge of biological science; others are the result of a narrow approach to insect control. Few studies have included basic investigations on the effects the chemicals might have on other components of the ecosystems to which the pests belong. The entomologist may recognize the desirability of a thorough investigation of these aspects, but because of the need for immediate answers to pressing problems and because of other pressures he does not have the necessary time.[27]

Moreover, in 1971, while reflecting on the controversy over multiple introductions, Carl Huffaker, Powers Messenger, and Paul DeBach responded to their Canadian critics by arguing firstly that although it was impossible to document the different degrees of control obtained with single and multiple introductions, there was no evidence at all that it was a counterproductive practice.[28] They also added more significantly that,

> Nicholson ... as former Chief of the Entomology Division of the Australian Council for Scientific and Industrial Research Organization ... did not use his own theoretical implication of a possible adverse effect, to deny any otherwise proper introductions

and that,

> it is entirely unrealistic to attempt to find and to pre-rank every possible candidate in order to ascertain the 'best' one to introduce ... The time (many years) and money spent to make such exhaustive studies before making any introductions would be imprudent and unnecessary.[29]

The roots of this attitude were recalled by DeBach, when he said, in 1987, that, "Smith never ... let theoretical work replace the goals of the Division of ... Biological Control ... and the Experiment Station, which were to solve pest problems".[30]

Yet, as Chant and Turnbull knew well, Smith's heirs justified the supposedly unsound practice of multiple introductions on widely accepted ecological grounds. The former had criticized this practice because Nicholson's theory suggested that it would reduce the effectiveness of the individual predators or parasites in controlling the density of the host or prey population, hence, the suggestion that entomologists should refrain from any introductions until the one most appropriate natural enemy was discovered. The entomologists from Berkeley and Riverside, on the other hand, rejected this argument, by pointing to what they thought was a highly successful record, and by then suggesting that this practice was perfectly consonant with the theory of "competitive displacement" first articulated by the Russian ecologist Georgii Gause.[31]

This theory of competitive displacement suggested that ecologically similar species could not coexist; that is, if two species were to compete for the same resource, the superior competitor would eventually drive the inferior one to extinction. According to DeBach, the history of the California Red Scale and its parasites illustrated why multiple introductions were not counterproductive, and also served to confirm the theories of competitive displacement elaborated by population ecologists. The Red Scale had been imported accidentally into California at the turn of the century; in order to check the damage it caused, biological control workers in California first introduced the parasite *Aphytis chrysomphali*.[32] In the late 1940s, its ecological homologue, *Aphytis lingnanensis*, was also introduced to California, in the hope of obtaining better control of the scale. Later still, a third homologue, *Aphytis melinus*, was introduced for the same purpose. By the mid-1960s, *Aphytis chrysomphali* had been virtually eliminated from California, because its range overlapped with those of the other two, superior parasites. These last two homologues, however, were found not to interfere with one another. The two species had found slightly different ecological niches, in different climatic regions of California, where each was the superior competitor. For DeBach and his associates this history of *Aphytis* was an obvious example of competitive displacement in action, and a reply to Turnbull and Chant's criticism about lack of theoretical grounding. More significantly, the interference between the competing species was transitory, and, therefore, unproblematic for the practice of biological control.

It was also possible, then, to view biological control practice from a perspective that was far less "theoretical" than that advocated by Turnbull and Chant. Smith and his followers viewed successive introductions of natural enemies as a number of ecological experiments, whose outcome would serve to establish guidelines for subsequent practice. If there then existed a body of theory that agreed with the lessons gained from these experiments, so much the better; such theory would lend more prestige to Smith's and his colleagues' standing as ecologists. And, if these lessons resolved some important difficulties in Nicholson's theory, they were quite substantial contributions to ecology. More significantly, all this suggests that the definition of what constituted integrated control was very open, and that closure around any one defintion would require much work of a different kind.

THE POLITICS OF BIOLOGICAL CONTROL IN RIVERSIDE AND BERKELEY

There can be little doubt that Albert Turnbull and Donald Chant's criticisms of work in biological control angered the entomologists in Berkeley and Riverside very considerably. In 1973, at the Summer Institute on Biological Control in Jackson, Mississippi, the former director of the Entomology Research Institute for Biological Control in Belleville, Bryan Beirne, rebuked the latter for their angry reactions by arguing that,

> to call a criticism an attack, and then to discredit the so-called attacker because he has attacked, does not bring credit to anybody. I suggest that it would be better for biological control in the long run to encourage criticisms and discuss them with professional

> objectivity than to try and frighten off potential critics. I suggest that arguments against criticisms are most constructive, most convincing, and most dignified when they are based on objective evaluation of facts. If the facts and arguments do not refute adequately, then perhaps there may be something valid in the criticism.[33]

Beirne assumed that "objective evaluation of the facts" would settle the disagreement between the entomologists at Riverside and Berkeley and those in Canada by either supporting or undermining their claims of about the effectiveness of multiple introductions and other controversial aspects of biological control practice. Yet, these "facts" were very ambiguous. Turnbull and Chant had shown that biological control did not work 100% of the time, but neither did it fail 100% of the time. Even worse, their estimate of the success rate of biological control programs was not substantially different from the figure published in 1964 by Paul DeBach, a figure which the latter considered a considerable achievement.[34] Chant has summarized this paradoxical situation by stating that, "the Californians thought in terms of 30% success rate (an achievement) whereas many of the rest of us thought in terms of a 70% failure rate, all more or less using the same sets of data".[35] Obviously, then, the disagreement over the effectiveness of classical biological control did not rest on empirical considerations alone.

Beirne himself explained to his audience in Jackson that the attitude of the entomologists in Berkeley and Riverside, as well as the "ad hoc" character of their biological control work, could only be attributed to constraints imposed by their practical responsibilities and the tremendous pressures on them to produce tangible results. He wrote that,

> a reason why criticisms of biological control procedures can produce emotional reactions is ... that administrators and entomologists in decision-making positions may notice the criticisms and be influenced by them to the disadvantage of biological control activities. This might be related to the fact that some of those who work in biological control are employed specifically as biological control specialists in agencies that have the word biological control in their titles. To preserve and promote the interests of their specialties and of their agencies they are in effect forced into a position of having to defend biological control against so-called attacks on it.[36]

There is much truth to Beirne's analysis. During the late 1960s, Smith's heirs had to confront some very serious institutional challenges, such as that from Roy Hansberry, a biologist in the Biological Sciences Research Center of the Shell Oil Company. In 1968, Hansberry argued on the pages of the *Bulletin of the Entomological Society of America* that,

> public concern over pesticides in the environment has stimulated research toward controlling insects without insecticides ... Few people realize the extent to which tax-supported work on proper use of insecticides has diminished. This drying up of insecticide research in experiment stations has happened while the use of insecticides has been increasing ... Yet this is just exactly the kind of research that has made the United States citizen the best fed person in the world, and it is just exactly the kind of

> research we need to keep food production even with people production ... This de-emphasis of chemical control research can be justified only if there are real prospects for success by indirect control methods even where non-insecticidal chemical are used. Otherwise agriculture and the public are being badly served.[37]

Hansberry then proceeded to review the data compiled by DeBach on the effectiveness of biological control, to conclude that it would never match that of chemical insecticides, and that its funding should be seriously reconsidered. This was a serious threat, for although the supporters of chemical control were confronting in the mid-1960s very serious public opposition to their approach — the movement to ban the use of DDT was gaining considerable momentum — they remained nonetheless very formidable adversaries. Not surprisingly, therefore, the entomologists in Riverside and Berkeley moved quickly, but, as we shall see in the next chapter, with great difficulty, to demonstrate that biological control was more effective than Hansberry would have had the taxpayer believe. In 1971 they responded by arguing that,

> Hansberry ... considered that the 61 cases of biological control for Canada and the USA is a paltry achievement. We contend that in view of the ... very limited financial support and extreme paucity of biological control specialists in the country, as contrasted to the very large monetary support and veritable army of workers in the pesticide area, this record is commendable ... For the amount of effort expended, we do not consider the number of successes to have been few — quite the contrary.[38]

Being fully aware of this difficult institutional situation, Beirne and other Canadian entomologists simply expected that the entomologists in Riverside and Berkeley would not tolerate any internal criticism of their accomplishments. Chant has described the situation writing that,

> biological control to the Californians was a religion with Harry Smith and Curtis Clausen (both wonderful gentlemen and scientists when they did not let their enthusiasm for biocontrol get in the way of their scientific objectivity) as its high priests and Paul DeBach, P.S. Messenger, Bob Van den Bosch, Ray Smith and others as their disciples. Anyone who questioned the faith, however obliquely, was a heretic.[39]

It was obvious to Chant and Turnbull that these entomologists in Berkeley and Riverside would try anything to improve the effectiveness of biological control. This, however, was only half the story.

BIOLOGICAL CONTROL AND ECOLOGICAL THEORY IN BELLEVILLE

Albert Turnbull and Donald Chant may have professed to believe that biological control,

> is sufficiently specialized to warrant consideration as a science in its own right, one that may draw heavily on other disciplines but no more. Its parts should be directed inwards to its own problems, not outwards to the problems of other subjects.

And that there should be an

> interplay between the search for principles and the solution of practical problems … Practical problems cannot be put aside indefinitely until we have accumulated enough information to attack them on purely rational grounds.[40]

Their preferred "principles", however, were not so completely unrelated to work in ecology. Turnbull and Chant's repeatedly voiced mistrust of the notion of a balance of nature seemed to be related to that skepticism about the ability of natural enemies to control the numbers of their victims often voiced by Robin Thompson from the 1930s onwards. For example, the two Canadian entomologists believed that the characteristics of the best natural enemy was a high rate of parasitization or predation, rather than a large area of search, as was recommended by John Nicholson. Thompson had emphasized this same characteristic for reasons ultimately connected with his belief that climate is a fundamentally important agent in shaping the dynamics of populations, and Chant, as we have seen, was similarly inclined to attribute great importance to this agent. There are, however, other, more direct reasons for believing that the divergent views of Chant, Turnbull and their counterparts in Riverside and Berkeley, on the effectiveness of biological control, were motivated by fundamentally different ecological views about the organization of biological communities and the regulation of the populations that comprised them.

In 1964, Chant had suggested that Ken Watt's mathematical modeling strategy provided a way to build better integrated control programs, but the latter's ideas were definitely grounded in ecological considerations. For example, in their definition of integrated control, Vernon Stern and his colleagues had followed Harry Smith by attributing the eruptions of insect species to the lack of a balanced natural enemy complex. This account was at odds with Watt's experiences while in the service of the Canada Department of Agriculture. During the late 1950s, he had been responsible for the development of ecological models from data collected during the famous "Green River Project", a long-term study of the population dynamics of the Spruce Budworm.[41] It was difficult for him to attribute the occasional, but dramatic outbreaks of the Budworm to any disruption of the balance of nature, since the Budworm was a native insect species on an equally native plant species. Furthermore, the pest was attacked by a large number of, again, equally native parasitic and predatory species, with no obvious stabilizing effect. According to Watt, the determination of Budworm densities could not be attributed, to the action of natural enemies alone. He then, undertook a detailed analysis of the whole environment of forest insects, not just of the interactions between them and their natural enemies, which led him, in turn, to question the generally accepted notion that the stability of ecological communities was correlated with their complexity.[42]

It was widely believed among population ecologists that the densities of animal species tended to be more constant in those communities that were constituted of greater numbers of animal and plant species. While the exact mechanism of stabilization was not fully understood, it was assumed that as the number of species increased, so would the number of checks on the growth of each one of them. The Spruce Budworm did not seem to fit the pattern. Watt suggested that the correlation of complexity and stability only held within, not across different levels of the food chain; in other words, the greater the number of natural enemies competing for a single prey or host species, the greater would be the interference between them. For example, a parasitic species might spend hours searching for an appropriate host within which to lay its eggs; if the chosen host had been parasitized, earlier, by some other species, none of the eggs would produce more parasites.[43] Under these circumstances, the densities of each parasitic species would be lowered, and perhaps stabilized. Thus, complexity assured some semblance of stability at this particular trophic level. However, each species' control over the density of the host would be less effective. Furthermore, under these conditions of diminished biological control, the densities of the species in the lower trophic level, the host, would become more sensitive to environmental variables than to the abundance of any one species of natural enemies. Complexity actually decreased stability at this lower level. These insights suggested to Watt that regulation and stability were functions of the whole environment of the preyed species; in his view, Nicholson and his followers had given an oversimplified account of the dynamics of populations. Needless to say, the introduction of multiple species of natural enemies was not to be recommended.

Such criticism of Nicholson's theory of population ecology was seconded by the second researcher cited by Chant, Watt's close associate in the Canadian Department of Agriculture, Buzz Holling. Holling's critique echoed many of the concerns expressed by Thompson on the relation between ecological and the physical models. Holling suggested on the basis of his experimental studies of Deer mice's predation on the European Pine Sawfly that Nicholson's approach, inspired as it was by the mathematical models of physicists such as his colleague, Victor Bailey, was inappropriate for the purposes of biologists, who studied systems far more complex than the ones encountered by physicists.[44] Holling wrote that, at one time, the models of the physicists,

> seemed to provide the population biologist with the same powerful tools that had proved so effective in analyzing processes in physics. The models attempted to express with great precision the action and interaction of a number of processes that affect animal numbers, and significant and important generalizations emerged from them concerning the gross behavior of populations. But if the ultimate test of usefulness of an approach lies in the power the resulting theories provide to manipulate events, these early applications of mathematics are not notably successful. It is still largely true that any animal population that affects man economically must still be approached as a unique entity. There are few principles that can be usefully exploited to manipulate a variety of situations.[45]

Holling was suggesting that the theory upon which the practices of the entomologists in Riverside and Berkeley were supposedly based had very little to say about the "real world", and therefore it could not possibly be of any real use to them. This is exactly how Turnbull and Chant viewed the situation, and it is not at all surprising that their own work informed, and was informed by, the ecological ideas of Holling and Watt.

Turnbull and Chant had read, in brief, the failure of an unsystematic approach from the very same data which had suggested to the entomologists in Riverside and Berkeley that the introduction of natural enemies to control insect pests was a highly successful practice. They attributed this failure to an excessive emphasis on practical accomplishment, an emphasis fostered by the insecure institutional status of the Division of Biological Control in the agricultural experiment stations at Riverside and Berkeley. However, they might well have read failure in this data even if the entomologists in Riverside and Berkeley had not been in such an institutionally precarious situation. Turnbull and Chant judged these practices as unsound not because they should not be grounded in ecological theory, but because they thought that the ecological theory supported by the entomologists in Riverside and Berkeley, Nicholson's theory, was unsound. It seems clear that Turnbull and Chant were much more committed to grounding practices of biological and integrated control systematically and directly in their own, preferred ecological theory. It should not be surprising that there were good institutional reasons for this orientation.

DIFFERENT INSTITUTIONS, DIFFERENT IDEAS ABOUT THEORY AND PRACTICE

As we saw in the two previous chapters, the Canadian federal government fostered, from the late 1940s onward, a rapid expansion of facilities for scientific research. Most of this support went to highly centralized scientific divisions in the various departments of the federal government, where it was administered by scientists who believed that the development of new technologies was crucially dependent on the prior advancement of knowledge about their theoretical foundations. Thus, in 1963, the Senior Entomologist in the Department of Agriculture, Edgar Le Roux, asked "how can we integrate chemical and biological agents successfully if we havn't the faintest idea of the fundamental role of these agents, singly or together, in the population dynamics of pest species?"[46] Furthermore, while discussing the pitiful quality of contemporary research on the effects of chemical insecticides, Le Roux added that,

> what is true for short-term pesticide research is by and large also true for short-term research on the effects of parasites and predators in control. Most of the results obtained on the performance of biological agents in population control are of the same order as those obtained for chemicals and for the most part misleading.[47]

For Le Roux, biological and integrated control would contribute to solving the problems caused by insecticides only if thoroughly grounded in long-term studies of insect ecology. This is exactly what Donald Chant, Buzz Holling, Albert Turnbull, and Ken Watt were trying to accomplish. Thus, while reflecting on these scientists' work, and on its relationship to that of the entomologists in Berkeley and Riverside, the former director of the Entomology Research Institute for Biological Control, Bryan Beirne, argued that they "were employed to do high-quality research rather than to solve individual pest problems ... [and] ... their individual work was much more fundamental than that of the Californians".[48] This "fundamental" work was, of course, ecological work. Notwithstanding their warning that "workers on long term projects frequently develop an attitude that leads them more and more to appreciate their work as an end in itself ... [and] ... will tend to drive this sort of project into basic ecology or pure biology", Chant, Holling, Turnbull, and Watt were focusing on the problems of biological and integrated control as if they were ones of insect ecology rather than economic entomology.[49] The Ecological Society of America certainly recognized Holling's work as such, when it awarded him with its prize for his work in ecology and acknowledged his debt to a very fortunate institutional situation. Moreover, federally supported research on chemical insecticides was affected just as profoundly by this orientation of Canadian agricultural research, and its share of the funding was not as overwhelming as in the United States. These scientists did not need, therefore, to affirm the great potential of biological control as constantly as their counterparts in Riverside and Berkeley; in fact, some of them supported, as these counterparts never could, an ecological theory which implied that biological control could not always work.

More significantly, Chant and Turnbull, like many other agricultural scientists in federal laboratories, were encouraged to adopt a "multi-disciplinary" and "mission-oriented" approach to their work. Since the development of integrated control required the cooperation of entomologists with very different expertise, for a very specific economic task, their work in this area fitted well with this policy. Chant and Turnbull were, however, most interested in general understanding of the variety of ecological factors, biological and otherwise, affecting the operation of various approaches to crop protection, rather than in translating this understanding into solutions for specific economic problems. Thus, Beirne complained that while Chant, Holling, Turnbull, and Watt's work "had some influence on some Canadian projects ... it did not, in my opinion, have any real impact ... or at least not yet".[50] Then, in the early 1960s, these scientists' situation became more difficult as ever greater stress was placed on universities rather than federal laboratories as the rightful centers for work on these more abstract dimensions of integrated control, and the need to suitably redefine the responsibilities of these federal laboratories. Chant was very worried about the consequences of this development for his research when he became the director of the more marginal Research Laboratory in Vineland. Under these circumstances, Chant and Turnbull (and many other colleagues in the Departments of Agriculture and Forestry) were faced with two choices. They could oppose the demands for greater economic relevance by pointing to the inadequacies and risks involved, or they could leave the laboratories for academic posts.

At first, Chant and Turnbull opted for the first strategy. They were well aware of the historical pressure on entomologists in Berkeley and Riverside to tailor their work to the needs of their agricultural constituencies, and that they disagreed over ecological matters. As we have seen in this chapter, they tied these two points together, in their assessment of biological control and its potential, to argue for the importance of ecological research to improve the effectiveness of both biological and integrated control. They suggested that less than a third of all introductions of natural enemies had succeeded to control the damage caused by the targeted insect pest. This, they then argued, was because the introductions often ran counter to the theoretical precepts supposedly informing them, and because the theory was, in any event, unsound. A more systematic approach to understanding the mechanics of biological control, in their view, readily revealed that natural enemies would not always control insect pests. This was not just a matter of insufficient institutional support for biological control, as the entomologists in Riverside and Berkeley claimed, but a matter of ecological constraints. Integrated control was therefore the way forward in crop protection, and a more ecologically sophisticated use of chemical insecticides would have to be centerpiece of the approach. This prospect was not recognized as such, Turnbull and Chant explained, wherever entomologists' institutional position was strongly dependent on success of biological control, as was often the case in Riverside and Berkeley. The implicit lesson was that the Department of Agriculture should not try to establish a similar organization of entomological research, and should continue instead to encourage the research tradition established during the 1950s.

Chant and Turnbull's was, however, a losing proposition. This was not because their analysis was problematic, but because the Canadian government was determined to transfer such research to the universities. They, and many other scientists in the Departments of Agriculture and Forestry then sought to move into more congenial academic positions in the United States, or into increasingly more attractive positions in Canadian universities. To accomplish this, they had to stress even more than before their work's contributions to debates far removed from the world of practical applications. This was one more reason, therefore, to stress the great ecological sophistication of their work. Once in their new, academic posts, they all began to focus on problems altogether different from the ones they had confronted while employed by the Departments of Agriculture and Forestry. In 1963, Watt left the Department of Agriculture for a post in the then newly established Institute of Ecology at the University of California in Davis. There, he became increasingly interested in problems of environmental and urban planning, and, by the late 1960s, he was hailed on the pages of *Time* magazine as one of the "New Jeremiahs", the prophets of environmental doom.[51] In 1964, Chant too moved to the United States. Quite ironically, he was appointed Director of the Division of Biological Control at Riverside. His mission there was to encourage a more theoretically sophisticated approach to the problems of biological and integrated control.[52] His plans were soon frustrated, and he moved once again, to the department of zoology at the University of Toronto. There he returned to his early work on the systematics of Spider Mites, and also became closely involved with the establishment of what is perhaps one of the most successful environmentalist groups in Canada, the Pollution Probe Foundation.[53] Holling stayed in the Division of Forest Biology of the Department of Forestry until 1967, when he became professor of zoology at the University of British Columbia. He continued work in population ecology, but, like Watt, he also became very interested in problems of environmental and urban planning. He too became a "New Jeremiah". Bryan Beirne followed a quite different path from any of these colleagues, which reflected the diversity of possible interests and commitments among entomologists employed by the Department of Agriculture. He took advantage of federal "development grants" to establish the first Canadian degree program in pest management at the newly created Simon Fraser University. Turnbull followed Beirne, but soon retired from the entomological scene to navigate around the world on merchant ships.

In the next chapter, I will turn our attention to the problems confronting the entomologists at Riverside and Berkeley as they, unlike their Canadian critics, sought to transform integrated control into something more than just an interesting idea about how to help the alfalfa and cotton growers of California — their careers depended on it. By the mid-1960s, they came around to the view that to resolve these problems, work of crop protection had to take into account the ecology of the entire crop and how it was affected by human action. The agricultural entomologist had to become, for them, some kind of agricultural manager on a par with industrial managers. Integrated control had to become integrated pest management. In the ensuing, sixth and last chapter, I will draw attention once again to ecological matters, and more specifically to debates during the late 1960s and early '70s over the ecological foundations of integrated pest management. As we shall see, those institutional factors which had characterized the conceptual development of biological control in Riverside and Berkeley during the 1950s and '60s, together with a new vision of the place of ecology in the world of American science, profoundly affected the theoretical work of Harry Smith's "grand-children".

NOTES

1 See S.E. Kingsland, *Modeling Nature. Episodes in the History of Population Ecology* (Chicago: University of Chicago Press, 1985), pp. 98–145.

2 A.P. Dodd, *The Biological Campaign against the Prickly-Pear* (Brisbane: Commonwealth Prickly Pear Board, 1940), pp. 1–13; and 107–46.

3 See A.J. Nicholson, "The self-adjustment of populations to change", *Cold Spring Harbor Symposia on Quantitative Biology* 22 (1957): 153–72; and also W. Kimler, "Advantage, adaptiveness, and evolutionary ecology", *Journal of the History of Biology* 19 (1986): 215–33.

4 A.J. Nicholson, "The balance of animal populations", *Journal of Animal Ecology*, 2 (1933): 132–78; and A.J. Nicholson & V.A. Bailey, "The balance of animal populations", *Proceedings of the Zoological Society of London* (1935): 551–98.

5 W.R. Thompson, "Biological control and the theories of the interactions of populations", *Parasitology* 31 (1939): 299–388.

6 Thompson, "Biological control", pp. 300–1.

7 L.O. Howard, *A History of Applied Entomology* (Washington: Smithsonian Institution, 1930), pp. 533–4.

8 See R. Doutt & P. De Bach, "Some biological control concepts and questions", in De Bach (ed.), *Biological Control of Insect Pests and Weeds* (New York: Reinhold Publishers, 1964), pp. 118–42.

9 H.S. Smith, "The role of biotic factors in the determination of population densities", *Journal of Economic Entomology* 28 (1935): 873–98.

10 P. De Bach & H.S. Smith, "Are population oscillations inherent in nature?" *Ecology* 22 (1941): 363–69.

11 Ken Hagen (University of California), interview with Paolo Palladino, 6 July 1988.

12 D.A. Chant, "Strategy and tactics of insect control", *Canadian Entomologist* 96 (1964): 182–201, on pp. 191–2. The references are to Stern *et al.*, "Integrated control"; R. Van den Bosch & V.M. Stern, "The integration of chemical and biological control of arthropod pests", *Annual Review of Entomology* 7 (1962): 367–86; and J.M. Franz, "The ecological effect of the control of insects by means of viruses and/or bacteria as compared with chemical control" in D.J. Kuenen (ed.), *The Ecological Effects of Biological and Chemical Control of Undesirable Plants and Animals* (Leiden: E.J. Brill, 1961), pp. 93–105.

13 See T.R. Dunlap, *DDT Scientists, Citizens, and Public Policy* (Princeton: Princeton University Press, 1981), p. 195.

14 Chant, "Strategy", p. 193.

15 Albert Turnbull to Paolo Palladino, 13 July 1992. The review was published as A.L. Turnbull & D.A. Chant, "The practice and theory of biological control of insects in Canada", *Canadian Journal of Zoology* 39 (1961): 697–751. See also Turnbull, "The ecological role of pest populations", *Proceedings of the Tall Timbers Conferences*, 1 (1969): 219–32.

16 See H.S. Smith, "Multiple parasitism: Its relation to the biological control of insect pests", *Bulletin of Entomological Research* 20 (1929): 141–9; see also R. Sawyer, "To make a spotless orange: biological control in California", (Ph.D. dissertation, University of Wisconsin, 1990), pp. 135–6.

17 Immediately after the second world war, Canadian entomologists established close connections, through the Commonwealth Institute of Biological Control, with those, such as Hans Zwölfer, who had been working on forestry insects in Germany and other central European entomological laboratories (see National Archives of Canada: record group 17: volume 3047: folder 400–6(1): J.J. DeGryse & W.E. Steenburgh, "Report on the biological control of spuce insects in central Europe", December 1946). They soon came to hold in very high regard studies such as Zwölfer' analysis of the effect of competitive interference in host-parasite complexes ("Untesuchungen über die Struktur von Parasitenkomplexen bei einigen Lepidopteren", *Zeitschrift fur Angewandte Entomologie* 51 (1963): 346–57).

18 A.L. Turnbull, "Population dynamics of exotic insects", *Bulletin of the Entomological Society of America*, 13 (1967): 333–7; cf. with P. De Bach, "The competitive displacement and coexistence principles", *Annual Review of Entomology* 11 (1966): 183–212; and R. Van den Bosch, "Comment on population dynamics of exotic insects", *Bulletin of the Entomological Society of America* 14 (1968): 112–5.

19 Turnbull & Chant, "Practice and theory", p. 744.

20 Turnbull & Chant, "Practice and theory", pp. 743–4.

21 Turnbull & Chant, "Practice and theory", p. 736.

22 Chant, "Strategies", p. 187.

23 Chant, "Strategies", p. 186.

24 Turnbull & Chant, "Practice and theory", p. 743.

25 Turnbull & Chant, "Practice and theory", p. 744.

26 Quoted in Turnbull & Chant, "Practice and theory", p. 744.

27 Stern *et al.*, "Integrated control", p. 85.

28 C.B. Huffaker & C.E. Kennett, "Studies of two parasites of olive scale, *Parlatoria oleae* (Colvee). IV. Biological Control of *Parlatoria oleae* (Colvee) through the compensatory action of two introduced parasites", *Hilgardia* 37 (1966): 283–335; and Huffaker *et al.*, "Natural enemy", p. 46.

29 C.B. Huffaker *et al.*, "The natural enemy component in natural control and the theory of biological control" in Huffaker (ed.), *Biological Control* (New York: Plenum Press, 1971), pp. 16–67, on p. 46.

30 Paul De Bach (now deceased) to Paolo Palladino, 9 July 1987.

31 See Kingsland, *Modeling Nature*, pp. 146–75.

32 See De Bach, "Competitive displacement".

33 B.P. Beirne, "Status of biological control procedures that involve parasites and predators" in F.G. Maxwell & F.A. Harris (eds), *Proceedings of the Summer Institute on Biological Control of Plant Insect and Disease* (Jackson: University of Mississippi, 1974), pp. 69–76, on p. 73.

34 See P. De Bach, "Successes, trends, and future possibilities" in De Bach, *Biological Control of Insect Pests and Weeds*, pp. 673–713.

35 Donald Chant (Ontario Waste Management Corporation) to Paolo Palladino, 5 January 1988.
36 Beirne, "Status of biological control procedures", p. 73.
37 R. Hansberry, "Prospect for nonchemical insect control — An industrial view", *Bulletin of the Entomological Society of America* 14 (1968): 229–35, on pp. 229–30.
38 Huffaker *et al.*, "Natural enemy", pp. 53–5.
39 Donald Chant (Ontario Waste Management Corporation) to Paolo Palladino, 5 January 1988.
40 Turnbull & Chant, "Practice and theory", p. 743.
41 See R.F. Morris (ed.), "The dynamics of epidemic spruce budworm populations", *Memoirs of the Entomological Society of Canada* 31 (1963): 3–332, and Watt's contributions to this volume.
42 K.E.F. Watt, "Community stability and the strategy of biological control", *Canadian Entomologist* 97 (1965): 887–95. See also D. Goodman, "The theory of diversity-stability relationships in ecology", *Quarterly Review of Biology* 50 (1975): 237–66.
43 Zwölfer, "Untersuchungen", pp. 354–6.
44 C.S. Holling, "The components of predation as revealed by a study of small mammal predation of the European pine sawfly", *Canadian Entomologist* 91 (1959): 293–320; see also "The functional response of predators to prey density and its role in mimicry and population regulation", *Memoirs of the Entomological Society of Canada* 45 (1965): 3–60.
45 Holling, "Functional response", p. 5.
46 E.J. Le Roux, "Ecological considerations in chemical control: insect population problems", *Bulletin of the Entomological Society of America* 10 (1964): 70–4, on p. 73.
47 Le Roux, "Ecological considerations", p. 73.
48 Bryan Beirne (Simon Fraser University) to Paolo Palladino, 12 February 1988.
49 Turnbull & Chant, "Practice and theory", pp. 743–4.
50 Bryan Beirne (Simon Fraser University) to Paolo Palladino, 12 February 1988.
51 "Ecology: the new Jeremiahs", *Time*, 15 August 1969, pp. 38–9.
52 Sawyer, *To Make a Spotless Orange*, p. 240.
53 See D.A. Chant (ed.), *Pollution Probe* (Toronto: New Press, 1970), pp. v–ix.

CHAPTER 5

Environmentalism and Entomological Research, 1960–1975

CONTROLLING PESTS AND IDEOLOGY

As we have seen in the previous chapters, DDT, which, during the 1940s, had seemed to many excited North American entomologists to usher a new age in the work of crop protection, was proving quite ineffectual just ten years after its first introduction. More alarmingly, the environmental disruption wrought by its widespread and ecologically indiscriminate use, polemically described by Rachel Carson in *Silent Spring*, resulted, during the 1960s, in a campaign in the United States and Canada to ban DDT.[1] Entomologists in the United States Department of Agriculture and in some of the largest state agricultural experiment stations, chiefly those in California, Texas, and North Carolina, responded to the situation by presenting integrated control and "sterile male release" programs, the release of artificially sterilized males to destroy the reproductive capacity of insect pest populations, as radically novel control programs that would eventually remedy the agricultural and environmental disaster for which the public held them responsible. During the late 1970s, however, entomologists were sharply divided over the merits of these two innovations. The historian John Perkins has argued that the division was so sharp because integrated control and sterile male release programs were based on views of the relationship between humanity and nature that were mutually incompatible.[2] Apparently, the federal entomologists supporting the sterile male release approach had no reservations about helping farmers by simply "eradicating" pest populations down to the very last individual because they believed human capacity to reshape the nautal environment to suit any material needs was unlimited. The proponents of integrated control, mostly from state agricultural experiment stations, argued instead that entomologists should limit themselves to "managing" the use of insecticides so as to limit any damage to the natural environment supposedly because humans were an integral part of nature and had no right to eliminate any other species. Given these opposing ideological commitments, it was perhaps inevitable that any discussion between the proponents of the two programs should result in sharp divisions with no room for compromise. This explanation, however, is at odds with the occasional willingness of some of the chief proponents of integrated control to argue that the two programs were quite compatible. Still other proponents seemed to oppose eradication programs on grounds that had far more to do with funding for entomological research than the unity of humanity and nature. In this chapter, I want to suggest that these contradictions are better understood by considering how the entomologists in the state agricultural experiment stations used, and combined, resources as disparate as this environmentalist ideology and the antagonism between federal and local authority that has characterized so much of the history of the United States, to advance their often precarious institutional situations.[3]

INSECTS AND CHEMICAL INSECTICIDES

By the 1950s, the unrestrained use of DDT, and other synthetic insecticides like it, was having tremendous repercussions on the ecology of insect species. In the short term, it resulted in the selection of resistant strains. In the longer term, the indiscriminate destruction of all insects, pests and their natural enemies alike, allowed any surviving insect populations to grow, or re-surge, quite explosively after each insecticidal treatment. Furthermore, species that once existed at naturally low densities because they were held in check by their natural enemies, and were therefore economically insignificant, became now more abundant; whereas they had formerly been treated as "secondary pests", they now became ones of primary economic importance. While farmers' response to this problematic situation was simply to increase the frequency and doses of the treatments, that of most entomologists was to develop new and more powerful insecticides. In so doing, resistance, resurgence, and secondary outbreaks were only worsened.[4] This was a cause of great concern to a number of North American entomologists, and, from the mid-1950s onward, they began to discuss the need for some better way out of this situation. Integrated control and the use of artificially sterilized males to eradicate insect populations seemed to be the most promising solutions.

Entomologists such as those in the Division of Biological Control at Riverside and Berkeley, who, following the advent of DDT, had found themselves relegated to the margins of the American entomology, moved back into the limelight. They injected new life into their programs by proposing the use of chemical insecticides in ways that would take into better account the ecology of noxious insect species and their natural enemies. For example, Vernon Stern, Ken Hagen, Ray Smith, and Robert Van den Bosch proposed to control the Alfalfa Aphid by spraying the crop with an insecticide, Systox, which was more lethal to the aphid than its major natural enemies.[5] The latter could then do most of the work in bringing the abundance of aphids down to an economically tolerable level. Yet, for all the rhetoric about the central role of such natural enemies in this integrated approach, in some ways, the latter was not fundamentally at odds with that of most contemporary entomologists. In the early 1960s, the entomologists in Riverside and Berkeley viewed the future of integrated control as contingent on the development of more, and better "narrow-spectrum" and "systemic" insecticides, a forecast which was in line with the well-established rule that the proper response to the failure of one class of chemical insecticides was to replace it with a new one. In another respect, however, integrated control was indeed based on a rather significant departure from current practices. Ever since the emergence of arsenical insecticides at the turn of the century, and even more so after the advent of the more effective organochlorine and organophosphorous insecticides, chemical treatment was on a fixed schedule, unrelated to the actual abundance of the targeted pests. The entomologists in Riverside and Berkeley sought to break with this tradition by arguing that control measures should be adopted when, and only when, the projected economic losses incurred by the growers exceeded the costs of treatment. Moreover, unlike entomologists of an earlier age (or many contemporaries, for that matter) their equation was not based simply on the cost of the insecticides against immediate gains. The value of the crop saved following the insecticidal treatment was supposed to be weighed against future damage due to resistance, resurgence, and secondary outbreaks. Under this broader set of considerations, these entomologists hoped that the attractiveness of insecticides would be severely dampened, and that biological control would eventually carry the day. This also implied, of course, that the goal of total freedom from insect pests should be abandoned as an unobtainable one.

There was at hand, however, a far more radical approach to crop protection, which promised to dispense eventually with all uses of insecticides; Rachel Carson described it in the following manner,

> about a quarter of a century ago Dr. Knipling startled his colleagues by proposing a unique method of insect control. If it were possible to sterilize and release large numbers of insects, he theorized, the sterilized males would, under certain conditions, compete wiith normal wild males so successfully that, after repeated releases, only infertile eggs would be produced and the population would die out.[6]

Edward Knipling, to whom Carson was referring, was the director of the Division of Entomology in the Agricultural Research Service of the United States Department of Agriculture from 1953 to 1970. From 1950s onward, he strove to shift the agenda for federally funded entomological research away from the single-minded emphasis on chemical control, which had by then resulted in the Department virtually terminating all work on biological control and other alternative practices. He sought renewed support for these alternatives and especially for the use of artificially sterilized males, an approach which he himself had been thinking about ever since the late 1930s. Knipling believed that if an insect population were so sparse that a female faced comparable odds of mating with either normal and fertile males or artificially sterilized ones, then it would be possible, after repeated releases of these sterile males, to destroy the reproductive capacity of the targeted population and eliminate it altogether; that is to say, it could be "eradicated". The Screwworm, which existed naturally at very low population densities, and yet was a cause of significant economic losses to the cattle industry in the southwestern United States, seemed to Knipling the perfect candidate for the approach. In 1954 and 1955, his idea was put to test on the Caribbean island of Curaçao, with remarkable success: after the treatment, Screwworms could no longer be found anywhere on the island. This approach was then used by the United States Department of Agriculture to eliminate the Screwworm from Texas and Florida, and in the early 1960s, Knipling suggested that perhaps it could also be used against the most intractable insect enemy of American agriculture, the Cotton Boll Weevil.[7]

ERADICATING THE COTTON BOLL WEEVIL

In the late 1960s, the United States Department of Agriculture obtained the financial backing of the National Cotton Council and political support in Congress to implement a Pilot Boll Weevil Eradication Experiment and test Edward Knipling's ideas on a grand scale. The experiment would cost, extraordinally, \$5.25 million, but, as cotton protection accounted for almost half the use of insecticides in the United States, and two thirds of them were used against the Boll Weevil alone, this truly enormous investment seemed worthwhile. If successful, the experiment would transform the sterile male release approach into the most revolutionary innovation in crop protection since the advent of DDT.[8]

The plan for the experiment was firstly to reduce the Boll Weevil population over a selected area to a very low density. Thus, starting in the fall of 1971, parts of Mississippi were subjected to a drastic control program, which included the destruction of crop residues at the end of each growing season to reduce the number of over-wintering refuges for the Weevils, intensive insecticidal treatments before and after the dormant season, and the use of hormonal attractants to trap most of the insects remaining after this preliminary phase. Artificially sterilized male Boll Weevils were then released to eliminate any survivors. As this trial run reached completion in 1973, many entomologists began to express serious doubts about the effectiveness of the method because a small number of Boll Weevils had been found in the interior of the experimental site.[9] A special panel of the Entomological Society of America was called to review the accomplishments of the experiment, and the panelists immediately sought to establish from where these Weevils had come. If they had simply survived the entomologists' onslaught, then the sterile male technique was far less effective than promised. The amazingly low Weevil densities achieved by this effort could only be maintained by continuously monitoring the crops and quickly treating local infestations. It was of course possible that the Weevils had simply migrated into the area after the onslaught. This second, seemingly less damaging explanation, however, drew attention to the discretely overlooked fact that, although the Screwworm had been completely eliminated from Curaçao, a long-term commitment of financial resources was needed in Texas to prevent any re-invasion of cleared areas. Such re-invasions might be particularly common in the case of the Boll Weevil, which was known to migrate over relatively long distances. Those entomologists in United States Department of Agriculture who had championed the revolutionary achievements of the sterile male release method were caught in a bind. Not only had they failed to fulfill the promise which had allowed them to obtain from their public and private sponsors previously unimaginable funding, they had also attracted the anger of entomologists outside the Department. It seemed to these latter entomologists, and especially to those advocating integrated control, that their colleagues in the Department of Agriculture had obtained their funding by simply misrepresenting the goals of their project, by promising to repeat the achievements on Curaçao, while knowing quite well that they were in fact aiming to obtain a situation similar to that established against the Screwworm in the Texas. They were not prepared to let the Department of Agriculture off the hook.

Five years after the Pilot Boll Weevil Eradication Experiment, the Entomological Society of America convened a round-table discussion on the merits of eradication programs.[10] The mediator, a senior administrator and entomologist in the Agricultural Research Service of the United States Department of Agriculture, H.C. Cox, opened the discussion by pointing out that even graduate students knew that, "eradication is a politicians' term. They really mean control".[11] Nevertheless, the North Carolina State University entomologist and articulate proponent of integrated control, Robert Rabb, took the term "eradication" to be something more than just a "politicians' term". He began his criticism by considering the different aims of integrated control and eradication. He did not believe that,

> man can devise large-scale productive and persistent agro-ecosystems in which he is the sole beneficiary of the energy and materials involved. The goal of pest management is not to eliminate pests, but rather to reduce their take of biomass as low as possible without jeopardizing the stability of production.[12]

For Rabb (and other proponents of integrated control), entomologists should aim to maximize the long-term productivity of farmers by balancing any benefits from the use of insecticides against any destruction of the pests' natural mechanisms. At times, this might even mean that no action should be the appropriate response. More crucially, however, Rabb argued that,

> the most rational general approach to pest problems is to learn to live with them … the use of [technological] power [against them] is a tremendous responsibility and must be done without arrogance and with subtle sensitivity, if not reverence, for the value of all life.[13]

These views of the aims of crop protection were sharply different from Knipling's. For Knipling, working in a very different institutional context, not dependent on the tenuous institutional fortunes of biological control, integrated control was nothing more than the systematic assembly of a number of different tactics to maximize the control of insect populations, while seeking to reduce dependence on chemical insecticides. The chief problem was to establish the proper balance between chemical control, whose efficiency is proportional to the density of the population targeted for destruction, and the release of sterilized males, whose effect on the target population is inversely proportional to its density. That the adoption of control measures should be contingent on economic deliberations of the type involved in the formulation of the economic thresholds advocated by Rabb and other supporters of integrated control was, for Knipling, absolute nonsense. In fact, he argued, firstly, that entomologists should strive to afford humanity with the "maximum protection from pests that affect [its] health and comfort", and then proceeded, secondly, to express his "great confidence in the ingenuity of our young scientists to perfect the technology necessary to put sound principles of insect suppression [eradication] into practice in future years".[14]

Knipling preferred, then, a terminal solution to an expensive co-existence with the insects feeding on American farmers' crops, and professed an unlimited confidence in the abilities of entomologists to obtain this goal. This was the expression of a more fundamental belief that humans' capacity to mold nature to suit their own needs was limitless and proper. Rabb disputed this faith on both points. Given such fundamentally incompatible perspectives on the interaction of humans and nature, it is not surprising that any disputes between the two groups over the relative merits of their respective programs should be intractable. There are good grounds, however, for suspecting that there was much more at stake than just ideological commitments.

For example, Paul De Bach, a member of that group of entomologists at Riverside and Berkeley who played a central role in the development of integrated control, was totally dedicated to the development of biological control as a self-standing strategy of pest control, and a long-time opponent of the aerial eradication campaigns so criticized by Rachel Carson (who had been helped by De Bach while writing *Silent Spring*).[15] Yet, like Carson, De Bach did not find the notion of totally eliminating an insect species particularly reprehensible, if it was accomplished by biological rather than chemical means.[16] In fact, he proposed to introduce insect species that were closely allied to ones attacking agricultural crops by way of their habits (except for the undesirable ones, such as feeding on these crops) and that were also superior exploiters of at least one shared resource. De Bach believed that the ensuing competition between these species would result in the displacement and extinction of the noxious ones. For yet another important and familiar figure in the early development of integrated control, the director of the Entomology Research Institute for Biological Control at Belleville, Bryan Beirne, eradication was one among many ecologically sustainable approaches to pest control; permanence, as opposed to the fleeting and very destructive action of insecticides, was for him the one goal toward which all entomologists should aim. In fact, in 1967 Beirne wrote that, "the most desirable level of results [of pest management practices] is permanent pest eradication".[17] In sum, De Bach's and Beirne's critical references to eradication were indictments of the indiscriminate use of insecticides rather than indictments of the ultimate goal: the complete elimination of some undesirable insects. Thus it seems that in 1960s integrated control and eradication were viewed by some entomologists as compatible approaches to crop protection.

More crucially, however, Carl Huffaker, the director of the Division of Biological Control at Berkeley, sharp critic of Knipling's approach, and one of the founders of integrated control, whose views provide most of the material for Perkins' contrast between the ideological foundations of sterile male release and integrated control programs, could sometimes accept the goal of eradication. In 1970, during a conference in Raleigh, North Carolina, on the "Principles of Pest Management", a conference which drew together the leading figures in the development of integrated control, Huffaker first asserted quite clearly that, "eradication is outside the philosophy of pest management by definition", but then went on to concede that the complete elimination of a pest population was, sometimes, a legitimate goal.[18] He suggested that small, recently established, and circumscribed populations, especially of species so destructive that management programs to control them and other secondary pests were effectively precluded, could be targeted for eradication.[19] These criteria — small, recently established, and circumscribed populations — were very vague. It is not at all clear why a program to eradicate the Woodwasp in Australia, which was described during the same conference at which Huffaker stated his views on eradication, satisfied them, while the Boll Weevil program did not.[20] Entomologists in the United States Department of Agriculture had often argued that the intensive insecticidal programs required to protect cotton from the Boll Weevil prevented the implementation of existing and effective management schemes to control a secondary, but very destructive insect, the Cotton Boll Worm.[21] Huffaker had suggested that in a situation such as this one, eradication of the more destructive pest might be a legitimate goal; yet, when it actually came to the Boll Weevil, he rejected this rationale.

Huffaker's inconsistent views raise some doubts about Perkins' account of the controversy. Confidence in it is eroded still further by the explicitly economic and professional motives for opposing eradication offered by some other participants in the crucial round-table discussion in 1978. An entomologist at Louisiana State University, advisor to Carson, and yet another central figure in the development of integrated control, Dale Newsom, was called upon to discuss the economic aspects of the controversy. He prefaced his position by stating that,

> I do not consider it morally wrong and ecologically disastrous to attempt to eradicate some pest species. To the contrary, I would argue that such attempts may be appropriate pest management strategies in some cases, worthy of substantial, continuing research effort. However, the accomplishments of eradication efforts have been so meager and the costs so great that an assessment of the eradication strategy is long

> overdue. In addition to the investments of manpower and money in eradication efforts by the [United States Department of Agriculture] ... substantial amounts are contributed by other agencies of the ... Federal Government, State agencies, grower organizations, and individual growers. Considering funds from all sources, the amount spent for eradication is much too high a percentage of the total available for pest control. When judged by the extremely high ratio of failure to success, application of the eradication principle is a poor investment in pest management.[22]

Thus, although he acknowledged that some entomologists objected to eradication on ethical grounds, Newsom himself did not view eradication and integrated control as mutually exclusive approaches. He was rather more concerned about the appropriation by the proponents of eradication of a large fraction of the total allocations for research on crop protection, by making promises which were for the most part unfulfilled. Ultimately, Newsom warned that the "wishful thinking" characteristic of eradication projects presented a threat to public support for all entomological research. He did not believe that,

> society can indefinitely continue to pour money and manpower into ill-conceived eradication programs that produce nothing but continued dangling of the carrot of eventual eradication before society. Eventually our profession will lose credibility.[23]

These concerns about the economic and professional implications of eradication programs were not of marginal importance. Indeed, in his comments during this round-table, Rabb himself pointed out that,

> there is a finite amount of resources (talent and money) to expend on entomological activities. How much of this is it prudent to expend on eradication efforts? ... And what will be the effect of expensive large-scale eradication programs on basic research?[24]

Such links between ideological, economic, and professional concerns suggest a much more complicated struggle than Perkins would have his readers believe, a struggle which can be understood more fully only by attending firstly to political changes which transformed the United States during the 1960s and early '70s, and by then examining how this transformation affected the institutional organization of American entomological research, especially the work of those entomologists who sought to develop integrated control into a major new approach to pest control.

SILENT SPRING AND THE FUNDING OF AGRICULTURAL RESEARCH

Prior to the publication of *Silent Spring*, the control of insects and the use of insecticides were matters of debate among a very small and exclusive circle, consisting primarily of growers, usually represented by the American Farm Bureau Federation, commodity groups like the powerful National Cotton Council, and the chemical industry.[25] These parties were extremely influential in shaping the policies of the United States Department of Agriculture because, until then, agriculture and agricultural policy had been an isolated domain which, by the 1960s, seemed to many urban observers to function according to very different rules from those in other areas of American politics and society.[26]

The chairman of congressional subcommittee overseeing budgetary appropriations for national agricultural policy between 1947 and 1992, Jaimie Whitten of Mississippi, for example, has been likened by the political scientist Christopher Bosso to a "feudal lord" because he has run agricultural policy according to his own priorities and those of his mostly cotton-growing constituency.[27] Thus, when the United States Department of Agriculture requested $25 000 to study new ways to control the Boll Weevil, Whitten rejected the request by appealing to the need for fiscal restraint and questioning the necessity for any further research on this problem because DDT did the job quite well (in 1966, he sought to resist the movement to ban DDT by writing *That We May Live*, an unashamed celebration of its marvels).[28] At the same time, however, he quickly approved a request by the same department for $1.1 million to build the Boll Weevil Research Laboratory. He did so only because the laboratory was to be located in his own congressional district; had the site been elsewhere the outcome would not have favored the plans of the Department of Agriculture.[29]

During the 1960s, this world of agricultural politics was radically transformed, as growers, commodity groups, the chemical industry, and elected officials like Whitten had to come to terms with a new actor, the environmentalist movement. The disastrous impact of the indiscriminate use of insecticides during the aerial eradication campaigns against pests such as the Gypsy Moth, Spruce Budworm, Elm Bark Beetle and, most notoriously, the Fire Ant, on bird and fish populations far removed from the areas targeted by these programs, resulted in considerable opposition to them among an increasingly numerous and prosperous suburban middle class.[30] The members of this middle class had very little understanding of those economic pressures motivating the widespread use of insecticides like DDT.[31] But, for all their opposition to what was happening on American farms, orchards, and forests, with all the implications it carried for the survival of the wildlife for which they cared so much, they were excluded from those political circles that controlled agricultural policy. This situation seemed to them profoundly undemocratic.[32] This disquiet was exacerbated by a more general concern: here, as with the radioactive radiation from testing of nuclear weapons, the government seemed unwilling to do anything to protect the public from an invisible, but ever-present, menace.[33]

In 1962, the President, John Kennedy, whose election in many ways epitomized the political triumph of this new sub-urban middle class, requested that the issues raised by Rachel Carson, in *Silent Spring*, be investigated by his

Science Advisory Committee. The very decision to turn to this body, rather than to the National Academy of Sciences, symbolized the fundamental change under way.[34] Although it had been usual practice to turn to the "Biology and Agriculture Board" of the Academy for technical and scientific advice about agricultural issues, this board was now seen as too closely allied to the interests of the agricultural establishment to provide a balanced assessment of the problem.[35] In fact its report, *Pest Control and Wildlife Relationships*, was strongly criticized for being unduly sympathetic to the interests of the agricultural establishment, and to those of the chemical industry in particular.[36] The panel on insecticides of the Science Advisory Committee, on the other hand, could hardly be said to represent the interests of this establishment, as only one of its nine members was affiliated with either the chemical industry or an agricultural institution, namely James Horsfall, from the state agricultural experiment station in Storrs, Connecticut. Apart from the differences in attitude and tone, however, the reports prepared by the two bodies agreed on the need to support more research on improving existing pest control strategies and to develop alternative ones. But this uncontroversial advice was bound to be without effect as long as funding for the research was not forthcoming from the congressional committees in charge of agricultural matters.

Even this situation began to change in the aftermath of *Silent Spring*, as the sudden burst of public attention to the previously uninteresting politics of agriculture accelerated the ongoing transfer of political power from rural to urban and suburban America.[37] Many legislators rallied behind the environmentalist banner, to challenge the power of established figures such as Whitten. For example, in 1963, the newly elected senator from Connecticut, Abraham Ribicoff, circumvented the agricultural committees by bringing the insecticide issue before a specially organized subcommittee of the Government Operations Committee, which then called on a wider range of witnesses than ever before, some representing interest groups that had previously been deliberately excluded from the debate.[38] In this changing atmosphere it became possible to obtain more federal funding for research on alternative pest control strategies. The budget for all entomological research in the United States Department of Agriculture was increased very substantially over the period 1964–65.[39]

The allocation of these new federal funds, however, affected the development of federal entomological research programs and those in the state agricultural experiment stations very differently. Until the early 1970s, when the United States Depart-ment of Agriculture underwent a radical reorganization, federal funding for agricultural research was channeled chiefly through the Department. Its Agricultural Research Service received about four-fifths of the allocation. These funds were then distributed in a manner such that the directors of the various divisions of the Service could exert extraordinary control over any areas of agricultural research that were not likely to be funded by the private sector or by the more fiscally conservative state legislatures.[40] It was thanks to this power that Edward Knipling could re-direct the federal research agenda for entomology toward the development of alternatives to chemical control programs. He also used this power to garner political and financial support for his program to eradicate the Screwworm.

The fifth of the federal funds for agricultural research remaining after the Agricultural Research Service got its share was divided between the fifty state agricultural research systems. This resulted in a very thin distribution of these funds to researchers outside the United States Department of Agriculture.[41] Scientists in the state agricultural experiment stations were forced, therefore, to rely largely on state legislatures and the private sector. The very practical demands of these sponsors resulted in entomological research that emphasized above all the improvement of chemical control. Thus, by the 1960s, entomological research programs in federal and state research centers were very different, with the latter hardly affected by Knipling's reforms.[42] This difference was exacerbated in the years following the publication of *Silent Spring*. More federal funds for research on alternative strategies of pest control were now available, but the largest share went to federal rather than state research programs. Furthermore, the largesse of the federal government was not matched by state legislatures in most agricultural states, as the environmental movement carried considerably less political clout there[43]. Thus, the very active, but quite exceptional research programs led by Carl Huffaker, Robert Rabb, and Perry Adkisson, in California, North Carolina, and Texas, programs which aimed to reduce agricultural producers' reliance on insecticides just as much as Knipling's, benefited very little from this increase.

SEARCHING FOR NEW SOURCES OF FUNDING

Given a political and institutional setting favoring the research agenda of scientists in the service of the United States Department of Agriculture, those in the state agricultural experiment stations who did not enjoy any substantial support beyond what was allocated to them by the state legislatures, had to seek support from agencies not usually associated with agricultural research. Thus, the National Association of Land-Grant and State Universities suggested, in 1964, that they should turn to the National Science Foundation, the National Institutes of Health, the Atomic Energy Commission, or the Department of Defense, for support.[44] Thus the pioneering work during the late 1950s by Vernon Stern and his colleagues in Riverside and Berkeley to develop an integrated control program for alfalfa growers and the dairy industry was supported by the National Institutes of Health, as was Rabb's research at North Carolina State University. But such diverse organizations could not support projects of a scale comparable to the Screwworm program, or the Pilot Boll Weevil Eradication Experiment. However, with the establishment of the International Biological Program, during the mid-1960s, the problem facing the proponents of integrated control seemed to be resolved.

The International Biological Program was conceived in the early 1960s by a number of biologists working in fields excluded from the post-war expansion of funding of the bio-medical sciences.[45] The American participants in this program argued widely that government support for their cause was necessary to fully develop the unique potential of ecology to "manage" the problems of environmental degradation that so alarmed the American public during this period. Following the publication of the 1965 President's Science Advisory Committee report, *Restoring the Quality of our Environment*, these ecologists were able to bring their case to the attention of the House of Representatives Subcommittee for Science, Research and Development.[46] As the historian Chunglin Kwa has argued, their rhetoric was so attuned to prevailing public and political concerns that, in 1970, the president, Richard Nixon, signed legislation allocating $40 million to the National Science Foundation, to support the quite overwhelming American contribution to the International Biological Program.[47]

Carl Huffaker, the former student of the ecologist Stanley Cain, chairman of the United States Committee for the International Biological Program, was an early participant in this endeavor.[48] Along with Bryan Beirne and others, he was involved in the development of an international, cooperative effort to study fundamental aspects of the biology and ecology of major insect pests.[49] However, following the American government's decision to massively fund the International Biological Program, this effort was overshadowed by a vastly larger project. Much in the spirit of systems ecology, the spirit guiding American involvement in the program, Huffaker felt that work in economic entomology made sense only as just one among many different aspects of crop production (albeit the most interesting one to insect ecologists), an enterprise which could be properly managed only by understanding the reciprocal interactions of large, complex agricultural ecosystems and human decisions regarding their use.[50] Thus he proposed to examine the structure of five separate "crop production systems" — alfalfa, citrus, cotton, pome fruits, and soybeans — and to then develop mathematical models, which, he and his colleagues hoped, would eventually serve to manage the use of chemical insecticides, biological agents, and cultural modifications, to control the various species attacking these crops. However, this grand plan to transform integrated control into "integrated pest management" ran into problems due to its explicitly agricultural character.

The rules governing the American side of the International Biological Program stipulated that, like all the proposals to the American National Committee of the International Biological Program, Huffaker's proposed project on "The principles, strategies, tactics of pest population regulation and control in major crop ecosystems" should be supported by the National Science Foundation.[51] The Foundation, however, was reluctant to back it since it viewed the project as agricultural, and, therefore, within the purview of the United States Department of Agriculture.[52] At this time, however, the relationship between the Department and the proponents of the integrated pest management project was rather tense. Huffaker had expressed reservations about the role it might want to play in the project, fearing that the Department would try to control its future development.[53] The latter was no less wary of its prospective partner. H.C. Cox, the future moderator of the round-table discussion on insect pest management and eradication, argued that the Department was already involved in programs that closely paralleled many of the activities contemplated by Huffaker and his colleagues. In fact, a number of projects relating to the ecological aspects of agricultural production, including crop protection, had been funded by the Department during the 1960s, and were later incorporated into International Biological Program, hoping to thus provide some coherence to this effort. The Department, therefore, viewed the integrated pest management project as nothing else but an attempt by state scientists to assert their institutional independence.[54] This wariness, if not antagonism, was further heightened by the failure of the leaders of the integrated pest management to invite Edward Knipling to the conference in Raleigh on the "Principles of Pest Management".[55]

Given this unhappy climate, and the reservations of the National Science Foundation, it seemed that Huffaker and his colleagues would be unable to obtain either political or financial support for their activities in the United States. This was particularly aggravating since the Food and Agriculture Organization of the United Nations, in 1965, in the wake of the successes of the scientists at the University of California in re-establishing cotton production in the Cañete valley of Peru, had endorsed what was still known as integrated control as one of the most promising answers to the problems of crop protection in the lesser developed countries.[56] Huffaker and his chief collaborators, Ray Smith and Perry Adkisson in particular, had to seek their support in the halls of Congress and the White House.

In the late 1960s and early '70s, environmental issues had become major items on the American political agenda. Both the Republican president, Richard Nixon, and the Democrats in the Congress were vying with one another to capture the support of the environmentalist movement, and "ecology" and "environment" became central items in the arsenal of political rhetoric. Furthermore, the Nixon Administration was aggressively pursuing a twin policy of decentralizing the powers of the federal government, and bringing about a greater degree of political accountability in all areas of governmental activity, including support of science.[57] At the outset, these developments enabled Huffaker and his colleagues to surmount the opposition of the National Sceience Foundation and United States Department of Agriculture to the national integrated pest management project. Adkisson successfully turned for support to his Texas congressman and powerful chairman of the House of Representatives Appropriations Committee, George Mahon. Smith instead turned with equal success to Nixon's chief of staff, John Erlichman, who then urged the National Science Foundation to support the integrated pest management project under the umbrella of the International Biological Program, and directed the United States Department of Agriculture to facilitate this operation.[58]

The long term effect of these political changes, however, was even more significant for the development of integrated pest management. In 1970, the competition to capture the vote of the environmental movement between Edmund Muskie, the Democratic senator from Maine who was preparing the ground for a presidential bid in 1972, and Nixon, resulted in congressional legislation and executive orders leading to the establishment of the Environmental Protection Agency and the Council of Environmental Quality.[59] Soon after their establishment, these organizations turned their attention to insecticides in the environment. Responding to the controversies surrounding DDT, the White House requested the Council of Environmental Quality to develop guidelines for amending the Federal Insecticide, Fungicide and Pesticide Act.[60] One of the most significant recommendations by the Council was the transfer of power to regulate the insecticide industry from the United States Department of Agriculture to the Environmental Protection Agency. While responding to public mistrust of the Department as a captive of the chemical industry, incapable, if not unwilling, to enforce existing regulations, this transfer of responsibilities was also part of the president's plan to weaken all independent federal centers of power, and the Department of Agriculture was foremost among them. A second, very important recommendation of Council of Environmental Quality, which had the same political effect, was to channel through the Environmental Protection Agency all federal support for research on "biologically integrated alternatives for pest control".[61] The definition of the legitimate beneficiaries of such support (written with the aid of Huffaker, Smith, Robert Van den Bosch, and Dale Newsom) was so worded that only integrated pest management programs could qualify. It laid great emphasis on the role of biological control and economic thresholds in these "biologically integrated" approaches to pest control.[62] Other techniques, central to eradication programs were given but a cursory treatment, and, although eradication was contemplated as one possible tactic of crop protection, its range of applications was very severely restricted.[63] Since eradication was a major component of the United States Department of Agriculture research brief, this definition of acceptable alternatives effectively excluded the Department from obtaining any new funds until it redefined its program. Not surprisingly, it was highly critical of the report.[64] More significantly, however, the Environmental Protection Agency was now competing directly with the Department of Agriculture for congressional allocations to cover the costs of funding environmental and related entomological research, thus diminishing the power the latter could exercise over the national research agenda.

Any mandates to address the environmental crisis were meaningless, however, in the absence of support for them in the congressional committees overseeing budgetary appropriations. The chairman of the House of Representatives Appropriations Committee, Mahon, and his colleague, Jaimie Whitten, the chairman of the Subcommittee for Agricultural Appropriations, both staunch supporters of the "iron triangle" linking their committees, the United States Department of Agriculture, and the various agricultural interest groups, were still in power, and they were not at all sympathetic to the intrusions of the environmental movement (represented by Council of Environmental Quality and the Environmental Protection Agency) into this domain.[65] They fought a bitter battle against the amendment of the Federal Insecticide, Fungicide and Pesticide Act. Mahon eventually handed all budgetary responsibilities for the new legislation over to Whitten, thus giving this staunch supporter of the pesticide industry partial oversight over the budget of the Environmental Protection Agency, even though one of the many reasons for calling on this new agency to control the development and uses of chemical insecticides had been to subvert the close alliance of the United States Department of Agriculture and the industry. This situation, in addition to Whitten's noted fiscal conservatism did not promise well for the development of "biologically integrated alternatives for pest control". Yet, if any program to support research on "ecological" approaches to pest control was going to receive any funds at all (as was demanded by the Nixon Administration), it was far more likely that congressional representatives from agricultural districts, such as Mahon and Whitten, would support the integrated pest management project rather than the one proposed by the United States Department of Agriculture. Their support for the Department of Agriculture was tempered by general suspicion of all federal programs as somehow interfering with the rights of individual citizens. The Pilot Boll Weevil Eradication Experiment was virtually designed to transform this suspicion into active opposition by requiring the coordination of agricultural practices of independent growers on a regional basis, and relying on a highly centralized organization and a great deal of coercion to do so. These measures seemed to violate the basic right of growers to do as they pleased on their own land. Integrated pest management, on the other hand, seemed to lend itself more easily to implementation on a farm-by-farm basis, and thus fit much better with prevailing political attitudes.[66] Furthermore, while the active role taken by the Administration in promoting integrated pest management, with all the accompanying environmentalist rhetoric, served very usefully to steal some of the thunder from Muskie's presidential bid, it did not ruffle unduly the feathers of the chemical industry. Insecticidal control was after all a central and permanent feature of integrated pest management, and the national integrated pest management project was, above all, a research program, nothing more.[67] Eventually, this combination of powerful (if reluctant) political support in Congress, and the environmentalist and "neo-Federalist" agenda of the Nixon Administration, provided for the smooth passage through Congress of the Federal Insecticide, Fungicide and Pesticide Act and the annexed line item mandating federal support for integrated pest management through both the Environmental Protection Agency and the United States Department of Agriculture.

THE END OF THE AGE OF AFFLUENCE

The alliances which had supported the new environmentalist agenda suddenly collapsed in 1973. In the wake of the "oil crisis" and the general downturn of the American economy, federal spending was severely curtailed and the legitimacy of this agenda now had to be judged in strictly economic terms.[68]

Scientific research was not immune to this changed environment. Carl Huffaker's integrated pest management project faced severe financial difficulties as the three agencies supporting it attempted to divest themselves of any responsibility for its continuation. The Environmental Protection Agency questioned the priopriety of its involvement, viewing the program as falling outside its scope of responsibilities; it was, after all, a regulatory agency. At most, the Agency was interested in supporting more research on "economic thresholds", research which would directly support its regulatory functions.[69] But any greater emphasis on such economic studies would achieve forced the National Science Foundation to acknowledge the fact that it was supporting research well outside its traditional domain.[70] In fact, with the termination of American support for the International Biological Program, following declining political interest and very strong criticism of its accomplishments, the National Science Foundation sought to end its involvement in the integrated pest management program. The United States Department of Agriculture, instead had been placed in an unusual situation when it was ordered to support Huffaker's project. Since, at the outset of the project, its leaders had feared that they might lose control over it if the Department became directly involved, they had sought to obtain this support only in the form of extra-mural research grants. However, this category of funds had been severely reduced by the Nixon Administration, as part of its campaign to reduce and restrict the power of federal agencies.[71] If forced to continue supporting Huffaker's program, the Department of Agriculture would have then had to provide direct logistical support and manpower for a research program over which it had minimal control, and do so out of funds designated solely for its own projects. The Department dug in its heels.[72] Huffaker and his colleagues found themselves having to cut $500,000 from their budget. The Environmental Protection Agency and National Science Foundation suggested that the cotton sub-project, which was directed by Perry Adkisson, should absorb the brunt of the cut, since "cotton research is funded from a variety of sources".[73] Yet, in 1973, the National Cotton Council, the most important organization of American cotton growers and most significant source of private funding for cotton research was already supporting the Pilot Boll Weevil Eradication Experiment, and was further committed to support the even more ambitious and expensive Boll Weevil Eradication Trial.[74] Adkisson was caught between federal agencies, which were cutting their appropriations, and the only alternative sponsor, the National Cotton Council, which was already in the pocket of the proponents of eradication. This was particularly infuriating since, as a member of the technical committee overseeing the Pilot Boll Weevil Eradication Experiment, Adkisson was well aware that the National Cotton Council was supporting a program that had failed to fulfill its ambitious promise to eliminate the Boll Weevil once and for all time, and that many influential members of the Entomological Society of America felt that something ought to be done about this.[75] What had been simmering discontent among the leaders of the integrated pest management project erupted into an open challenge of the United States Department of Agriculture research program in entomology and crop protection. Once again, the prevailing political environment provided them with an opportunity to challenge the federal establishment with great effectiveness.

In the early 1970s, the entire research program of the United States Department of Agriculture had come under intense scrutiny. Up until that time, state agricultural experiment stations were apportioned federal funds for research in proportion to the population of each state.[76] The White House sought to eliminate such "formula funding" so as to reduce the power granted thereby to the Department of Agriculture. A controversial report by the National Academy of Sciences provided a cover for this political campaign. The *Report of the Committee on Research Advisory to the United States Department of Agriculture*, otherwise known as the "Pound Report", suggested that funding by formula led to excessive duplication of efforts and low quality, if not "pedestrian", research, and that a competitive grant system would be far more effective.[77] This condemnation was soon followed by another study by the National Academy of Sciences of current pest control strategies and prospects for the development alternative ones.[78] The chairman of the new investigative team, Donald Kennedy from Stanford University, was very sympathetic to the findings and recommendations of the "Pound Report". Adkisson figured very prominently in this project as a member of the Executive Committee, and so did the future historian of integrated pest management, John Perkins, in the role of Principal Staff Officer.[79] The committee's final report, *Pest Control: An Assessment of Present and Alternative Technologies*, appeared in 1975, after three years of exhaustive investigations. It was quite critical of United States Department of Agriculture research programs in the area in general, and in particular, it was extremely critical of the Boll Weevil eradication program. The Executive Committee complained that,

> the agricultural research enterprise now lacks adequate capacity to conduct experiments on the large scale necessary to evaluate the impact of changes in practice on a particular crop over a wide area ... The only projects of such scale to have drawn significant support to date, unfortunately, are eradication programs.[80]

The authors of the report then went on to endorse integrated pest management resoundingly, striving to emphasize the fundamental differences between eradication programs and integrated pest management.[81] Finally, the Committee expressed the belief that,

> the promise offered by this general approach [integrated pest management] makes it an appropriate candidate for high-priority attention by federal funding agencies.[82]

All this left the researchers in the United States Department of Agriculture very embittered because it seemed to imply that they were wholly opposed to integrated pest management, even though during the preparation of the report they had pointed out repeatedly how they had made significant contributions to the development of alternative approaches to chemical control that were both integrated and biological, and had also argued that the merits of the Boll Weevil eradication program had been judged unfairly.[83]

Notwithstanding the protestations of the Department of Agriculture, the National Academy of Sciences report was very influential in the White House Office of Science and Technology and the Office of Management and Budget, and thus had an important impact in redirecting the efforts of the United States Department of Agriculture away from eradication, and toward greater support for integrated pest management.[84] Although the Department of Agriculture still managed to obtain funds during the mid-1970s to proceed with its Boll Weevil Eradication Trial, such funding was conditional on support for an Optimum Pest Management Trial.[85] This last program was essentially identical to the one developed by Adkisson in Texas during the 1960s and further refined in the context of the integrated pest management program headed by Huffaker.[86] But the tide was running even more ominously against the supporters of eradication programs.

Although the first report by the Council of Environmental Quality on problems of pest control, prepared with the advice of the chief proponents of integrated pest management, recommended that control programs should not aim to completely eliminate any insect population, even eradication in the classical sense — the wide cast, usually aerial, use of chemical insecticides — was still considered a sometimes acceptable approach. This relatively catholic attitude was completely absent from the Council's second report on integrated pest management, even though it was written with the aid of virtually the same entomologists responsible for the first one.[87] It portrayed a far more antagonistic relationship between eradication and integrated pest management; for example, the first principle defining the last strategy was that,

> potentially harmful species will continue to exist at tolerable levels of abundance. The philosophy is to manage rather than to eradicate the pests.[88]

Here eradication referred specifically to the Pilot Boll Weevil Eradication Experiment, not any classical spraying program.[89] This second report by the Council of Environmental Quality served as the basis for president Jimmy Carter's directive to the relevant federal agencies, chiefly the Department of Agriculture, to "modify as soon as possible their existing pest management research, control, education, and assistance programs and to support and adopt IPM [integrated pest management] strategies wherever practicable".[90] The proponents of integrated pest management appeared to have won the day; they had marshaled a far more powerful coalition than the Department of Agriculture ever could, and were then able to force the Department to adopt their own agenda for pest control. The age of "I(mperial) P(otomac) M(adness)" the widely used acronym for integrated pest management, had begun.[91]

INSTITUTIONS, POLITICS, AND ENVIRONMENTALIST IDEOLOGY

In sum, the arrangements for funding of American agricultural research, up to the 1960s, resulted in a tense relationship between researchers in the United States Department of Agriculture and in the state agricultural experiment stations, and especially between the entomologists in these two organizations who were working to replace existing practices of pest control with economically more sustainable and environmentally less disruptive ones. The mounting concern of the American public, during the 1960s, over a seemingly inarrestable destruction of their natural environment, relieved some of this tension as more funds were made available for any research aiming to halt such destruction. But, later in the decade, this atmosphere of relative conviviality began to vanish, and it finally disappeared in the mid-1970s, as the American economy began to falter. Those researchers in the state agricultural experiment stations of California, Texas, and North Carolina, who led a national program for the development of integrated control into integrated pest management, were able to shore up their program by drawing some very sharp distinctions between what they were doing and the eradication program promoted by the Department of Agriculture. By drawing attention to a particularly popular view of the relationship between humanity and nature that seemed compatible with integrated pest management alone, a view upon which John Perkins built his account of the rivalry between the two groups of entomologists, the proponents of integrated pest management were able to enroll on their side an ally previously shared with the researchers in the Department of Agriculture, namely the environmentalist movement. But this move could not have succeeded if the appeal to environmentalist values, to distinguish the two programs, had not attracted the support of the Nixon Administration. The Administration was already predisposed to support the national program for the development of integrated control into integrated pest management as a ploy to both undermine the power of the United States Department of Agriculture, and to seek out the support of the environmentalist movement. The appeal to environmentalist values served to cement very strongly the alliance between the Administration and the proponents of integrated pest management. This account in terms of political alliances, however, does not necessarily mean that all the appeals to environmentalist values were motivated by political opportunism alone. It is quite likely that some of the proponents of integrated control and integrated pest management, such as Robert Rabb, genuinely believed that their strategy and eradication programs were predicated on fundamentally different conceptions of the relationship between humanity and nature. Carl Huffaker and Ray Smith certainly expressed some reservations (albeit very ambiguously) about the legitimacy of eradication in writings for their fellow supporters of integrated control well before the 1970s. However, the importance of these reservations must be tempered by considering H.C. Cox's view that no one truly believed in idea of eradication — the complete elimination of an insect population or species — and the more pragmatic, institutional concerns voiced by such important figures as Dale Newsom.

That the ideological differences between the chief proponents of integrated pest management and eradication programs by themselves were of limited importance to the long-term evolution of more ecologically sophisticated and less environmentally disruptive pest control programs becomes rather clear by taking stock of the achievements of the two programs by the late 1970s. Eradication was not a "once and for all" solution to pest problems, as had been promised before Congress. The Pilot Boll Weevil Eradication Experiment and Boll Weevil Eradication Trial were in practice long-term programs based on the use of multiple tactics, and the exploitation of crucial features of the reproductive ecology of insect populations. The divergence between promised and effective achievements had been the main source of criticism by the Entomological Society of America and National Academy of Sciences panels. The term eradication was eventually dropped in favor of the more cumbersome, but more accurate term "area-wide population management". Was integrated pest management any different? The notion of economic thresholds certainly lent distinctiveness to the approach developed by the proponents of integrated pest management. Yet, it was widely acknowledged that it was very difficult to determine what variables should be taken into account in the economic decision making model, and that this problem was aggravated by the lack of critical knowledge of population dynamics. Under such circumstances, building reliable integrated ecological and economic models that would permit the estimation of these thresholds was more an ideal than a practical reality.[92] Much more research in environmental economics, bio-mathematics, and insect ecology was needed, and it is precisely to the latter two areas of research which I will turn in the next chapter. All that remained of integrated control and integrated pest management was a program based on the use of multiple tactics, and greater consideration of the ecology of insect populations.[93]

NOTES

1 T.R. Dunlap, *DDT. Scientists, Citizens, and Public Policy* (Princeton: Princeton University Press, 1982).

2 J.H. Perkins, *Insects, Experts and the Insecticide Crisis. The Quest for New Pest Management Strategies* (New York: Plenum Press, 1982).

3 On the meaning of "ideology", see T. Eagleton, *Ideology* (London: Verso, 1991).

4 See R. Doutt & R.F. Smith, "The pesticide syndrome. Diagnosis and suggested prophylaxis" in C.B. Huffaker (ed.), *Biological Control* (New York: Plenum Press, 1971), pp. 3–15.

5 V.M. Stern *et al.*, "The integrated control concept", *Hilgardia* 29 (1959): 81–101.

6 R. Carson, *Silent Spring* (Boston: Houghton Mifflin Company, 1987), p. 279.
7 See Perkins, *Insects*, pp. 97–126; and E.F. Knipling, "Potentialities and progress in the development of chemosterilants for insect control", *Journal of Economic Entomology*, 55 (1962): 782–6.
8 E.F. Knipling, "Boll Weevil and pink bollworm eradication: progress and plans", *Ginners' Journal & Yearbook* (1971): 23–30.
9 W.G. Eden *et al.*, "The Boll Weevil Eradication Experiment", *Bulletin of the Entomological Society of America* 19 (1973): 218–21.
10 Entomological Society of America, "Eradication of plant pests — Pro's and con's", *Bulletin of the Entomological Society of America* 24 (1) (1978): 35–54.
11 H.C. Cox, "Comments of the moderator", *Bulletin of the Entomological Society of America* 24 (1978): 35.
12 R.L. Rabb, "Eradication of plant pests — Con", *Bulletin of the Entomological Society of America* 24 (1978): 40–4 on p. 43.
13 Rabb, "Eradication", pp. 40 and 44.
14 E.F. Knipling, "Eradication of plant pests — Pro", *Bulletin of the Entomological Society of America* 24 (1978): 44–52, on pp. 45 and 51.
15 See Dunlap, *DDT*, p. 194.
16 For Carson's opposition to the use of insecticides to eradicate insects, see *Silent Spring*, pp. 155–72; and on her attitude toward the total elimination of insect species by any other means, see pp. 277–91 (especially pp. 284–5).
17 B.P. Beirne, *Pest Management* (London: Leonard Hill, 1967), p. 71.
18 C.B. Huffaker, "Summary of a pest management conference: a critique" in R.L. Rabb & F.E. Guthrie (eds), *Concepts of Pest Management* (Raleigh: North Carolina State University, 1971), pp. 227–42, on p. 232.
19 Huffaker, "Summary", pp. 232–3.
20 See P.W. Geier, "Temporary suppression, population management, or eradication: how and when to choose" in Rabb & Guthrie, *Concepts*, pp. 170–90.
21 Richard Ridgway (United States Department of Agriculture), interview with Paolo Palladino, 20 June 1988.
22 L.D. Newsom, "Eradication of plant pests — Con", *Bulletin of the Entomological Society of America* 24 (1978): 35–40 on pp. 36 and 38.
23 Newsom, "Eradication", p. 39.
24 Rabb, "Eradication", p. 43.
25 For an analysis of the agricultural policy-making process in the 1950s and '60s, see C.J. Bosso, *Pesticides & Politics. The Life Cycle of a Public Issue* (Pittsburgh: University of Pittsburgh Press, 1987), pp. 21–78; and D.F. Hadwiger, *The Politics of Agricultural Research* (Lincoln: University of Nebraska Press, 1982), pp. 115–68.
26 See A. Mayer & J. Mayer, "Agriculture, the island empire", *Daedalus* 103 (1974): 83–95.
27 Bosso, *Pesticides*, pp. 65–71.
28 Hadwiger, *Politics of Research*, pp. 162–7; and Bosso, *Pesticides*, p. 130.

29 Perkins, *Insects*, pp. 114–5; and Bosso, *Pesticides*, p. 67.

30 Bosso, *Pesticides*, pp. 85–90; and Dunlap, *DDT*, pp. 89–90.

31 For a review of the politics of growth and the environmentalist movement, focusing especially on the class composition, see J.R. Logan & H.L. Molotch, *Urban Fortunes: The Political Economy of Place* (Berkeley: University of California Press, 1987), pp. 200–247, especially 209–223. See also S.P. Hays, *Beauty, Health, and Permanence: Environmental Politics in the United States* (Cambridge: Cambridge University Press, 1987), pp. 287–297.

32 Carson, *Silent Spring*, p. 127; with regard to the general political aims of *Silent Spring*, see D. Fleming, "Roots of the new conservation movement", *Perspectives in American History* 6 (1972): 7–91, on pp. 28–34.

33 R.H. Lutts, "Chemical fallout: Rachel Carson's *Silent Spring*, radioactive fallout, and the environmental movement", *Environmental Review* 9 (1985): 211–225.

34 President's Science Advisory Committee, *Use of Pesticides* (Washington: United States Government Printing Office, 1963). See also Bosso, *Pesticides*, pp. 117–8. On environmentalist agenda of the Kennedy Administration see H.P. Caulfield, "The conservation and environmental movements: An historical analysis" in J.P. Lester (ed.) *Environmental Politics and Policy: Theories and Evidence* (Durham: Duke University Press, 1989), pp. 13–56 (especially pp. 26–48).

35 Archives of the National Academy of Sciences: Biology & Agriculture: Committee on Pest Control and Wildlife Relationships: 1962–1963. The total budget for the study was $31,000, of which total, $10,750 was contributed by the pesticide industry, $15,500 by government agencies, and the rest by other interest groups.

36 See Dunlap, *DDT*, p. 116; for a broader assessment of this and other related National Academy of Sciences studies, see P.M. Boffey, *The Brain Bank of America. An Inquiry into the Politics of Science* (New York: McGraw-Hill, 1975), pp. 199–226.

37 Bosso, *Pesticides*, p. 125.

38 F. Graham, *Since Silent Spring* (Boston: Houghton Mifflin, 1970), pp. 74–5 and 176–81.

39 Perkins, *Insects*, p. 36, and Bosso, *Pesticides*, p. 121.

40 For illuminating reactions of the Entomological Society of America to the reorganization of the Agricultural Research Service of the United States Department of Agriculture during the 1970s, see F.W. Maxwell *et al.*, "Report of the committee on impact of ARS-USDA reorganization on the profession of entomology", *Bulletin of the Entomological Society of America* 20 (1974): 50–53; and K.L. Knight *et al.*, "Report of the 1975 ESA committee on impact of ARS-USDA reorganization on the profession of entomology", *Bulletin of the Entomological Society of America* 22 (1976): 65–73.

41 See E.L. Peterson, *The Research Program of the USDA*, Statement before the Federal Council on Science and Technology, April 28, 1959; L.E. Hawkins *et al.*, *Fund Support to the State Agricultural Experiment Stations*, Statement before the Division of Agriculture, Association of State Universities and Land-Grant Colleges, November, 1964; and United States Department of Agriculture, *Pesticide Program for Fiscal Year 1965* (Washington: United States Government Printing Office, 1965).

42 United States Senate, Hearings before the Subcommittee on Reorganization and International Organizations, 88th Congress, Second Session, *Inter-Agency Coordination in Environmental Hazards* (Washington: United States Printing Office, 1964).

43 For a review of the relationship between federal and state responses to environmentalists' concerns, see C.E. Davis & J.P. Lester, "Federalism and envrionmental policy" in Lester, *Environmental Politics and Policy*, pp. 57–84.

44 L.E. Hawkins *et al.*, *Fund Support to the State Agricultural Experiment Stations*, Statement before the Division of Agriculture, Association of State Universities and Land-Grant Colleges, November 1964.

45 See C. Kwa, *Mimicking Nature: The Development of Systems Ecology in the United States, 1950–1975* (Ph.D. dissertation, University of Amsterdam, 1989). On the precarious scientific status of American ecology which motivated much of their effort to establish the International Biological Program, see C.H. Waddington, "The origin" in E.B. Worthington, *The Evolution of IBP* (Cambridge: Cambridge University Press, 1975), pp. 4–11, on p. 9.

46 President's Science Advisory Committee, *Restoring the Quality of our Environment* (Washington: United States Government Printing Office, 1965). This report was prepared at the request of the Johnson Administration, in response to the political seachange expedited by the publication of *Silent Spring*. See E.J. Burger, *Science at the White House* (Baltimore: Johns Hopkins University Press, 1980), p. 76: and W.H. Lambright, *Presidential Management of Science and Technology* (Austin: University of Texas Press, 1985), pp. 44–48.

47 C. Kwa, "Representations of nature mediating between ecology and science policy: the case of the International Biological Programme", *Social Studies of Science* 17 (1987): 413–42.

48 Fred Smith (University of Michigan), interview with Paolo Palladino, 8 September 1988; see also "Section UM to contact the Commission de Lutte Biologique and the Commonwealth Bureau of Biological Control", *IBP News* 1 (1964): 27.

49 See D.F. Waterhouse, "Definition and planning of the project" in V. De Lucchi (ed.), *Studies in Biological Control* (Cambridge: Cambridge University Press, 1976), pp. 7–9.

50 Carl Huffaker (University of California), interview with Paolo Palladino, 15 August 1988. For a brief account of the development of Huffaker's plan

in the context of International Biological Program, see the annual reports of the United States National Committee for the International Biological Program, *Research Programs Constituting United States Participation in the International Biological Program* published by the National Academy of Sciences between 1968 and 1974. See also C.B. Huffaker & R.F. Smith, "The IBP program on the strategies and tactics of pest management", *Proceedings of the Tall Timber Conferences* 4 (1972): 219–236.

51 John Neuhold (Utah State University) to Paolo Palladino, 17 June 1988. Neuhold was, at the time, the Director of Ecosystem Analysis Program in the National Science Foundation.

52 Terry Davies (Conservation Foundation) and William Dickinson (Environmental Protection Agency), interviews with Paolo Palladino, 22 June 1988; Warren Muir (private consulting firm), interview with Paolo Palladino, 20 July 1988. Davies and Dickinson were, at the time, in the White House Office of Management and Budget; Muir served, instead, on the Council of Environmental Quality.

53 Files of Waldemar Klassen (United States Department of Agriculture): NSF-IBP/RANN-Dr. Huffaker: C.B. Huffaker to Steering Committee, October 14, 1971; Minutes of Steering Committee Meeting, December 2, 1971.

54 Klassen Files: NSF-IBP/RANN-Dr. Huffaker: H.C. Cox to T.W. Edminster, September 27, 1971.

55 See R.L. Rabb, "Book review of *Insects, Experts, and the Insecticide Crisis. The Quest for New Pest Management Strategies*", *Agriculture and Human Values* 1 (1984): 34–7.

56 Food and Agriculture Organization, *Report of the FAO Symposium in Integrated Pest Control* (Rome: Food and Agriculture Organization, 1966).

57 See M.D. Reagan & J.G. Sanzone, *The New Federalism* (New York: Oxford University Press, 1981), pp. 3–29; J.E. Katz, *Presidential Politics and Science Policy* (New York: Praeger, 1978), 180–8; and S.D. Fries, "The ideology of science during the Nixon years, 1970–1976", *Social Studies of Science* 14 (1984): 323–41.

58 Carl Huffaker (University of California), interview with Paolo Palladino, 15 August 1988.

59 Bosso, *Pesticides*, pp. 151–4; and E.J. Burger, *Science at the White House* (Baltimore: Johns Hopkins University Press, 1980), pp. 89–90.

60 Bosso, *Pesticides*, pp. 143–77.

61 Council on Environmental Quality, *Integrated Pest Management* (Washington: United States Government Printing Office, 1972).

62 Council of Environmental Quality, *Integrated Pest Management*, p. vii.

63 Council of Environmental Quality, *Integrated Pest Management*, p. 7.

64 Richard Ridgway (United States Department of Agriculture), interview with Paolo Palladino, 20 June 1988.

65 Bosso, *Pesticides*, pp. 6 and 154.

66 Waldemar Klassen (United States Department of Agriculture), interview with Paolo Palladino, 24 June 1988; and Robert Rabb (North Carolina State University), interview with Paolo Palladino, 30 June 1988.
67 William Dickinson (Environmental Protection Agency), interview with Paolo Palladino, 22 June 1988.
68 Bosso, *Pesticides*, pp. 174 and 178–81.
69 Klassen Files: NSF-IBP/RANN-Dr. Huffaker: W. Klassen to R.J. McCracken, 30 October 1973. See also National Academy of Sciences-National Research Council, *Pest Control: An Assessment of Present and Alternative Technologies*, Vol. I, *Contemporary Pest Control Practices and Prospects: The Report of the Executive Committee* (Washington: National Academy of Sciences, 1975), pp. 168–169.
70 Klassen Files: NSF-IBP/RANN-Dr. Huffaker: C.H Hoffman to H.O. Graumann, 26 September 1973.
71 Waldemar Klassen (United States Department of Agriculture), interview with Paolo Palladino, 24 June 1988.
72 Klassen Files: NSF-IBP/RANN-Dr. Huffaker: W. Klassen to R.J. McCracken, 30 October 1973.
73 Klassen Files: NSF-IBP/RANN-Dr. Huffaker: W. Klassen to R.J. McCracken, 30 October 1973, p. 2.
74 Perkins, *Insects*, pp. 139–41.
75 Perkins, *Insects*, pp. 127–39.
76 See V.W. Ruttan, *Agricultural Research Policy* (Minneapolis: University of Minnesota Press, 1982), pp. 81–82, and 255–256.
77 See National Academy of Sciences-National Research Council, *Report of the Commitee on Research Advisory to the United States Department of Agriculture* (Washington: National Academy of Sciences, 1972). For reviews and critical assessments of the report, see N. Wade, "Agriculture: NAS panel charges inept management, poor research", *Science* 179 (1973): 45–7; "Agriculture: critics find research stunted and wilting", *Science* 180 (1973): 390–91; and R. Levins, "Fundamental and applied research in agriculture", *Science*, 181 (1973): 523–4.
78 National Academy of Sciences, *Assessment*.
79 J.H. Perkins, "Strategies for pest control: The NAS report", *Proceedings of the Agricultural Research Institute* 28 (1979): 105–121; John Perkins (Evergreen College), interview with Paolo Palladino, 11 June 1988; Donald Kennedy (Stanford University), interview with Paolo Palladino, 15 September 1988. From the very outset, the committee responsible for this study was viewed as biased against the "agricultural establishment." National Academy of Sciences Archives. A: C&B: ESB: X: 1971–1973. CNR/ESB: Problems of Pest Control: Executive Committee, 1972: General: J.H. Perkins to R.A. Carpenter, 28 August 1972.
80 National Academy of Sciences, *Assessment*, p. 18.

81 National Academy of Sciences, *Assessment*, pp. 380–99.

82 National Academy of Sciences, *Assessment*, p. 386.

83 Richard Riley (United States Department of Agriculture), interview with Paolo Palladino, 21 June 1988. See also National Academy of Sciences Archives: A&C: CNR: ESB: X: 1973–1976: CNR/ESB: Problems of Pest Control: Review Process: 1975: E.F. Knipling to R.E. Green, 29 April 1975; E.F. Knipling to P. Handler, 24 June 1975; and R.W. Long to T. Shad, 15 July 1975.

84 Phil Smith (National Academy of Sciences), interview with Paolo Palladino, 16 September 1988; Donald Kennedy (Stanford University), interview with Paolo Palladino, 15 September 1988.

85 R.L. Ridgway & E.P. Lloyd, "Evolution of cotton insect management in the United States", and J.L. Hamer *et al.*, "Optimum Pest Management Trial in Mississippi" in Ridgway *et al.* (eds), *Cotton Insect Management with Special Reference to the Boll Weevil* (Washington: United States Government Printing Office, 1983), pp. 3–27 and 385–407.

86 D.R. Rummel *et al.*, "An appraisal of a 10-year effort to prevent the westward spread of the boll weevil", *Bulletin of the Entomological Society of America*, 21 (1975): 6–11.

87 This report was in large part the work of Dale Bottrell, a close associate of both Huffaker and Adkisson; Huffaker (University of California), interview with Paolo Palladino, 18 August 1988.

88 Council on Environmental Quality, *Integrated Pest Management* (Washington: United States Government Printing Office, 1979), p. vi.

89 Council on Environmental Quality, *Integrated Pest Management* [1979], p. 105.

90 J. Carter quoted in Council on Environmental Quality, *Integrated Pest Management* [1979], p. 116.

91 R.W. Stark, "Integrated pest management: back to the caves", *Bulletin of the Entomological Society of Canada* 10 (3) (1978): 56–9, on p. 56.

92 See Klassen Files: NSF-IBP/RANN-Dr. Huffaker: H. Hoffman to H.O. Graumann, September 26, 1973; as well as Council on Environmental Quality, *Integrated Pest Management* [1979], pp. 21–3; and B.A. McCarl, *Economics of Integrated Pest Management*, Special Report 636 (Agricultural Experiment Station, International Plant Protection Center, and the Department of Agricultural and Resource Economics: Oregon State University, 1981).

93 David Pimentel (Cornell University), interview with Paolo Palladino, 3 August 1988.

CHAPTER 6

Ecological Modeling and Integrated Pest Management

LINKING INTEGRATED PEST MANAGEMENT AND ECOLOGY

By the early 1970s, integrated control was transformed institutionally into integrated pest management, and had come to be widely regarded as the most promising approach to the protection of agricultural crops. For many American entomologists, this transformation also changed their professional image. No longer were they pedestrian tinkerers in agricultural science; they were instead the harbingers of a fundamentally new and technologically advanced approach to agriculture. Unlike traditional work of improving chemical insecticides and their modes of application on field crops, developing integrated pest management programs promised to be something like a feat of engineering. The main task of the future crop manager, just as in engineering, would be to find, with the aid of the most sophisticated technologies and computing equipment, optimal solutions to problems involving conflicting needs: how should one manage the use of chemical insecticides so as to maximize economic returns to farmers, while also minimizing those ecological disruptions that all too often undermined agricultural productivity over any term longer than one growing season? The answer to such a problem, the proponents of integrated pest management promised, rested on the application of the most recent developments in bio-mathematics and theoretical ecology. Yet, just as with earlier definitions of integrated control, the exact relationship between ecology and integrated pest management was far from clear.

In 1971, at the symposium in Raleigh, North Carolina, on "Concepts of Pest Management", Robert Rabb stated that,

> pest management deals primarily with populations, communities and ecosystems. Thus the basic biological discipline involved is unequivocally ecology, and there should be no fuzzy thinking about this fact.[1]

Yet, in a later symposium on the relationship between ecology and integrated pest management, which was held in 1984 at the meeting of the Entomological Society of America in San Antonio, Texas, another entomologist who also played an important role in the development of integrated pest management, Marcos Kogan from the University of Illinois, asserted that,

> the 1960s were a period of ecological awakening ... [integrated pest management] was called applied ecology, and the connection between ecology and integrated pest management was simply taken for granted. In reality, however, only species ecology (life histories and phenologies) and elementary population dynamics were requisite for the design, and even the implementation, of some of the most successful integrated pest management programs ... integrated pest management lacks a solid theoretical background is a matter of grave concern.[2]

It is clear from these remarks that the relationship between ecology and integrated pest management was not as simple and straightforward as Rabb had suggested. Some important roots of this difficulty can be understood by considering three important perspectives on ecology, and more specifically, three competing views on the role of modeling in ecology, which emerged and blossomed during the 1960s. The existence of these three competing views prevented, and still prevents, the formulation of any unambiguous statement about the nature of ecology. Since each of these alternative views was also influential in the development of integrated pest management, the definition of the relation between ecology and integrated pest management must consequently be an equally difficult problem.

ECOSYSTEMS ECOLOGY

The development of integrated pest management was contemporaneous with that of a novel perspective in ecology, "systems ecology". During the 1960s, the proponents of this perspective, thanks to their extensive connections with the Atomic Energy Commission, came to dominate the discipline, and intended to transform the latter's place in the political economy of American science. Ecology would no longer be the purview of wildlife managers and fishery biologists, but was to become an exciting and rigorous science.[3] This movement to transform ecology was the result of integrating two relatively distinct schools of thought.[4] On the one hand, it was strongly influenced by the work of the brothers Eugene and Howard Odum, located respectively at the University of Georgia and at the University of Florida. Their path-breaking studies of nutrient cycles and ecological energetics, such as the cycling of radio-nuclides through food chains on the Eniwetok atoll after it was blasted by a series of nuclear weapons tests, were grounded quite explicitly in an organicist world-view.[5] On the other hand, it was also heavily indebted to the work in population ecology by Ken Watt, in the Statistical Research Service of the Canadian Department of

Agriculture.[6] Starting in the late 1950s, Watt sought to deploy the new mathematical tools he had learned to use in another Cold War institution, the Rand Corporation, to model the behavior of the Spruce Budworm in the forests of northern Ontario. His approach would become in the following years one of the hallmarks of systems ecology.

In his extremely popular textbook, *Fundamentals of Ecology*, which drew together for the first time animal and plant ecology, Eugene Odum argued that the ecosystem was the fundamental unit of ecology, and he took the lake as its exemplar.[7] He emphasized the complexity of the network of reciprocal, balancing interactions connecting the innumerable biological and physical components of the lake, and that the transfer, storage, and recycling of nutrients and energy through this network bound the components together in a quite inextricable manner. Furthermore, he stressed that the dynamic phenomena characterizing such a network were contingent on the assembly of species and their physical environment into a specific organization of the food chain. In other words, these phenomena were no longer apparent when the component elements of the network were isolated from one another. Odum claimed, therefore, that these dynamic phenomena were irreducible properties of the ecosystem and that the latter had to be regarded as an autonomous, self-regulating unit.[8] Not surprisingly, the highly individualistic views entrenched in the works on the ecology of animal populations by, for example, John Nicholson and Robin Thompson, found little place in *Fundamentals of Ecology*; moreover, when Odum took some interest in population ecology during the debates during the 1970s and '80s over the "units" or targets of natural selection, he supported the more holistic accounts of group selection.[9]

Howard Odum attempted to articulate more fully and then formalize his brother's notions about the irreducible nature of ecosystems by relying on analogies with electric circuits. He believed that by taking these analogies very seriously, it would be possible to formulate more general statements about this unique nature of ecosystems. If ecologists were to adopt this approach, he went on to argue in *Environment, Power, and Society*, ecology could finally break away from the old, deplorable natural history, and finally become a rigorous science on a par with physics.[10] However, this "New Ecology", as Eugene Odum called it, would fundamentally differ from physics insofar as it would be a science of irreducibly integrated systems.

The Odum brothers, and Howard especially, also believed that, like the "New Physics" which produced the atomic bomb and nuclear energy, the development of this "New Ecology" would provide humanity with a powerful tool for the rational utilization of natural resources. Howard's *Environment, Power, and Society* was an explicitly technocratic manifesto about how to solve the environmental crisis facing humanity. The two brothers knew, however, that those recent, exciting developments in physics and engineering, would have been impossible without a close collaboration between physicists, engineers, and mathematicians. Thus, to complete the parallel, the further advancement of the intellectual revolution and of the ecological engineering which they proposed would require the development of appropriate mathematical tools, mathematical tools which would capture the unique properties of ecosystems.[11]

MATHEMATICS AND POPULATION ECOLOGY

In the chapter on biological control and ecology, we began to see how the 1940s and '50s were important years for the movement to formulate ideas about the dynamics of animal populations in the language of mathematics. From its very origins some thirty years earlier, this movement was inspired by the work of mathematical physicists, and in fact many of its members have been physicists or mathematicians turned ecologists.[12] Analogies with molecular kinetics proved fruitful for clarifying the nature and form of the competitive interactions between members of a single species, and between species, interactions which presumably regulated the numbers and distribution of animals in nature. These population ecologists shared two fundamental assumptions about nature, which were first articulated most clearly by the senior entomologist in the Australian Commonwealth Scientific and Industrial Research Organization, John Nicholson.

Reflecting on the famous control of the prickly pear by *Cactoblastics cactorum*, and the results of his studies of the Blowfly — a major problem for Australian sheep farmers — Nicholson argued more abstractly that the density of a population had to be regulated by factors whose effective power increased in strict proportion to it, for only such a factor could maintain the size of the population in equilibrium with its resources. According to Nicholson, competition for these limited resources, predation, and parasitism were the factors that acted in such a manner. He considered the effect of physical factors, and climatic ones in particular, on such a population to be unrelated to its density. While such factors determined the maximum capacity of the environment to support any given population, they could not affect the dynamics of that population when its density was changed from its proper one. Corollaries of this theory often seemed to be contradicted by the observed behavior of natural populations. However, those articulate proponents of the Nicholsonian view of the dynamics of natural populations, the entomologists in the Division of Biological Control at Riverside and Berkeley, and Harry Smith, Paul De Bach, and Carl Huffaker in particular, provided much experimental evidence in its favor.

The assertion that physical factors played a qualitatively distinct and secondary role in determining the number and distribution of animals was, however, very controversial. Although Nicholson's debate with Robin Thompson died out during the late 1930s, the issues it raised resurfaced fifteen years later, when Nicholson and his followers were challenged by two Australian zoologists, Herbert Andrewartha and Charles Birch. They started a sometimes bitter polemic, which supposedly reached a climax in 1957, at the prestigious Cold Spring Harbor Symposium on Quantitative Biology, when the geneticist and evolutionist Theodozius Dobzhansky confessed to having felt "for several years now that this is a controversy chiefly about words".[13] There must have been, however, more substantial differences between the two competing views, for the debate continued into the 1960s and '70s: the controversy between Donald Chant and Albert Turnbull, on the one hand, and the entomologists in the Division of Biological Control, on the other, was just one of its more recent manifestations.

Some of the issues at stake in this long-standing debate were made most evident when the competing ecological perspectives were expressed in mathematical form. Those population ecologists, who followed in the tradition set by Nicholson, could develop elegant mathematical models based on a very limited number of variables, which supposedly captured the essential features of the regulatory processes. Their critics regarded these models as appropriate only to describe the relatively simple systems studied by physicists. For example, Andrewartha and Birch argued that the distinction drawn by physical scientists between first order and second order interactions was justified on empirical grounds. In the case of a falling body, the effect of gravitational attraction was sharply distinguishable from the deflecting and slowing effects of the wind and air friction. But the distinction drawn by Nicholson and his followers between "density dependent" biological factors, and "density independent" physical factors, was not equally well supported by field data. For example, Andrewartha's observations during the 1930s of the population dynamics of the Australian Locust taught him that climatic factors could sometimes act as density dependent factors: the greater the population density, the greater the chances that some individuals would be left with no refuge against climatic mortality factors. Thus mortality imposed by these physical factors would be proportional to the population density. Andrewartha and Birch's later studies of *Thrips imaginis* confirmed this reasoning, a reasoning which seemed to strike at the heart of Nicholsonian ecology, challenging the very fundamental assumption that there existed a categorical distinction between density dependent and density independent factors.

In other words, Andrewartha and Birch viewed the Nicholsonian school as sacrificing realism, defined as close matching of model and empirical evidence, for the sake of elegance. However, Andrewartha suggested that the work of Ken Watt seemed to show the way toward a modeling strategy that would transcend the crude abstractions of Nicholsonian ecology.[14]

FROM SYSTEMS ANALYSIS AND POPULATION ECOLOGY TO SYSTEMS ECOLOGY

Ken Watt, who had been encouraged by his mentor in the Canadian Department of Agriculture, Thomas Burnett, to study biometry and population ecology under Thomas Park at the University of Chicago — then the major American center for the study of population biology — was one of the major figures in the development of a new approach to population ecology which aimed to break out of the impasse made so evident at the symposium in Cold Spring Harbor. Before proceeding any further, a clarification is needed on the use I be will making of the term "theoretical population ecologist": I will use it to identify those ecologists who advocated the view that animal populations are regulated in a Nicholsonian fashion, even though many of them were not nearly as inclined toward mathematical abstraction as Nicholson; the simpler term "population ecologist" is, however, inadequate because many critics of this abstraction, whose views I will discuss in the course of the chapter, also identified themselves as "population ecologists".

Beginning with his earliest experimental studies of the Flour Beetle, Watt came to mistrust the deductive models developed by theoretical population ecologists such as Nicholson, because they appeared to be based on assumptions which he thought were simply unrealistic. For him, physical factors, and accidents of place and time, were fundamentally important in determining the numbers and distributions of animal populations.[15] However, Watt also mistrusted purely descriptive models based on linear regressions, such as those developed by scientists in the Canadian Fisheries Research Board, which he came across as biometrician in the Statistical Service of the Canadian Department of Agriculture. He was just as suspicious of the very similar models developed during the "Green River Project" by his future colleagues in the Forest Insect Division.[16] They all seemed to Watt to be of limited utility to understanding the underlying causes of biological processes.[17] He suggested that a compromise between these two modeling approaches would be far more fruitful.

During the symposium on "Concepts of Pest Management" in Raleigh, Watt recalled how surprised he had been by the counter-intuitive information about pest control practices, which he and his colleagues in Forest Insects Division had gathered during the late 1950s and early '60s as they sought to understand the population dynamics of the Spruce Budworm in Canadian forests.[18] They found, for example, that the use of chemical insecticide to control the Budworm resulted in a greater frequency of outbreaks, and greater costs to the forestry industry. Watt believed that such counter-intuitive situations were quite common in ecology because so many interdependent variables were involved in any particular ecological interaction. Such complicated phenomena, he argued, were simply intractable within the traditional analytical framework adopted by theoretical population ecologists. A special approach would be needed to explain such anomalous situations as the failures of insecticides against the Spruce Budworm.[19] The philosophy of "operations research" and the language of "cybernetics", which he had learned after a stay at the Rand Corporation, to work with Richard Bellman, seemed to provide a better way of thinking about these problems.[20]

Operations research had been developed during the 1940s and '50s by military strategists and industrial engineers intent on obtaining optimal solutions to the problems of coordinating large scale military operations and industrial production involving complex feed-back loops. The aim of their work was to understand how variations in the relationships between the diverse components of the relevant operations affected the efficiency of military performance or industrial output. Such work also allowed these military and industrial researchers to identify the most critical components, which could then be manipulated to improve efficiency.[21]

Watt proposed in the *Annual Review of Entomology* that entomologists and ecologists should adopt this style of research, reversing the order of investigation.[22] In operations research, the components of the system and their mutual interrelations were known ahead of time, and the goal was to manipulate these components so that the system might work more effectively towards its prescribed end. System analysts, on the other hand, started from the final output, and sought to discover how it came about; that is to say, they were interested in discovering the structural organization of the system. Insect ecologists interested in formulating meaningful and mathematically rigorous statements about their subjects, Watt argued, should adopt this last strategy and become systems ecologists.

After leaving the Canadian Department of Forestry in 1963 for a post in the Institute of Ecology in the University of California at Davis, to work on problems of environmental and urban planning, Watt dedicated himself to a more general and systematic development of this mathematical approach to ecology. Thus, in his 1966 book, *Systems Analysis in Ecology*, Watt argued that the components of the ecosystem and the links between them were so numerous and varied, that an analytical model of it, one which started by building from prespecified principal causes, would quickly prove quite intractable. Simplification would perhaps make it less so, but it would inevitably lead to a loss of precious information. Such a model would therefore be very abstract and unrealistic, that is to say, it would fail to produce a close match with the observed phenomena.[23] Population ecologists, according to Watt, should therefore begin with the observed phenomena, for example, with the density of insect populations, as optimal solutions to an unknown organizational structure. Their first step should then be to formulate a structural model relating all the variables presumed to be implicated in the determination of this density; at this stage the various variables should be treated as black boxes, as their specific characteristics were of secondary importance to their quantitative effect on population density. The second step was to determine from data gathered in the field these effects, over some specified time-span. A statistical analysis would then allow both the identification of the variables most responsible for the observed outcome, and the consequent simplification of the initially very complex model. These steps were then followed by the formulation of mathematical rules defining the numerical relationships between the remaining variables. Finally, ecologists would analyze these mathematical rules, attempting to define their meaning in terms of biological phenomena. This could be done by starting a new cycle of black boxing, calculation, and analysis. The ultimate aim was to obtain a nested set of mathematical expressions which accurately described the dynamics of the insect population in question. Watt, of course, recognized that this approach was very time-consuming, but with the emergence of increasingly powerful computers, and better programming languages, such as FORTRAN, he believed that the application of systems analysis would inevitably transform ecology.[24]

More importantly, however, Watt's approach to ecological problems started with data about the higher units of organization, data about the ecosystem level, rather than with assumptions about fundamental interactions of species or populations. It was, in many ways, holistic, and was, therefore, very appealing to ecologists such as Eugene and Howard Odum. It seemed to provide them with a powerful tool for the mathematical formulation of their insight into the irreducible nature of ecosystems.[25] The unsympathetic ecologist Robert McIntosh suggested that what followed was a "shotgun marriage" between systems analysis and ecosystems ecology, which then gave birth to illegitimate systems ecology.[26] Significantly, this intellectual alliance was oiled by Watt and Eugene Odum's collaboration to establish a national Institute of Ecology dedicated to understanding how to forestall the impending environmental "Doomstime". As Watt told Odum,

> if ecologists do not get involved up to their necks in these problems, fast, they may be treated to the spectacle of watching executive and legislative branches of government appoint aerospace firms to do the job, simply because the public has confidence in the ability of the aerospace people to solve tough, large problems in a hurry. I am prepared to help ...[27]

Watt's fears, as we shall see, were not unfounded.

SYSTEMS ECOLOGY AND INTEGRATED PEST MANAGEMENT

There is a striking similarity between the ecological ideas articulated by ecosystems ecologists and those of the earliest proponents of integrated control. Like ecosystems ecologists, they stressed the importance of viewing pest control problems from an "ecosystem" perspective. For example, in their first formal definition of integrated control, the entomologists at the University of California in Riverside and Berkeley, Vernon Stern, Ray Smith, Ken Hagen and Robert Van den Bosch, stressed the importance of basing alternative pest control practices on a careful consideration of the ecosystem, which they defined as,

> the interacting system comprised of all living organisms of an area and their non-living environment. The size of area must be extended enough to permit the paths and rates of exchange of matter and energy which are characteristic of any ecosystem.[28]

Such language was typical of ecosystem ecologists' rhetorical repertoire. Moreover, the balance of insect populations within the ecosystem, Stern and his colleagues went on to argue, could be effectively manipulated only by taking into account “the oneness of any environment, natural or man-made”.[29] It is not at all surprising then that, in 1962, Van den Bosch and Stern acknowledge their intellectual indebtedness to Eugene Odum’s *Fundamentals of Ecology*.[30] Such emphasis on ecosystems was not unique to Stern and his colleagues, as they were joining a long list of entomologists advocating such a perspective. The real problem, however, was how to translate it into practical prescriptions about how to control insect populations. It was no easy task to predict how variations of the physical, biological, and social variables characterizing agricultural ecosystems might affect the activities of key insect species. In 1966, the Canadian entomologist Donald Chant, who two years earlier had become the newest director of the Department of Biological Control at Riverside, suggested that systems theory would provide the tools for this purpose, and he referred his audience to the pioneering work of his former colleague, Ken Watt.[31] Tying these new approaches to the ecology of insect pests and new mathematical tools together, and making integrated control into more than interesting idea, required new forms of institutional support.

As we saw in the last chapter, these new forms of institutional support emerged during the mid-1960s, with the establishment of the International Biological Program. Responding to widespread popular concern during the 1960s about the deterioration of the environment, the ecologists calling for this program argued that a new approach to the management of natural resources was required, and that systems ecology fit the bill. What was needed, they said, was “Big Biology” to match the evolution of the physicists and engineers’ “Big Science”, and the “New Ecology” would play the role of the “New Physics”.[32]

Paradoxically, the case for systems ecology and the International Biological Program was presented to the United States Congress by Fred Smith, a theoretical population ecologist, from the University of Michigan, who strongly supported the Nicholsonian theories of population ecology. In fact he was most famous for papers, which he co-authored with Nelson Hairston and Larry Slobodkin, in which he asked “why is the world green?” and answered by making a case for the fundamental importance of competition and predation in the regulation of natural populations.[33] Nevertheless, Smith argued that systems analysis would provide ecologists with a new perspective on problems that had proved intractable within the traditional deductive framework, and concluded his case by suggesting that the further development of this perspective would lead to a more effective application of ecology to human welfare.[34] In fact, the International Biological Program Report for 1974 was quite explicit about this managerial role for systems ecology, asserting that “this research program in ecosystem operations is aimed at the ultimate understanding of the environment and how manmade changes will affect it”.[35] These claims seem to have been very impressive, because Congress voted to contribute $40 million to the International Biological Program.[36] This moment was hailed by many ecologists sympathetic to systems ecology as marking a watershed in the history of ecology; the theoretical population ecologist Daniel Simberloff, on the other hand, has decried it as the most expensive and wasteful ecological program ever, perhaps because it was "wholly conceived in a holistic, ecosystem vein".[37]

Carl Huffaker, the director of the Division of Biological Control at Berkeley, was one of the earliest participants in the organization of the International Biological Program — he had been drawn into it by his former advisor Stanley Cain — and was also the driving force behind the establishment of an ecologically based pest control research program within its framework. Following the style set by systems ecologists, Huffaker proposed to enlist entomologists, agronomists, economists, and systems analysts from a number of American universities to examine the structural similarities between five different agricultural ecosystems. Confirming Ken Watt's worst fears, many of these systems analysts were recruited from the National Air and Space Administration as the Apollo Project came to an end after the landing on the Moon in 1969.[38] All these scientists' work, Huffaker argued, would eventually provide general rules for a more efficient management of agricultural resources.[39]

The proponents of integrated control had, in sum, emphasized both the fundamental unity of ecosystems, and the importance of systems analysis to understand the behavior of the different agricultural ecosystems under different control regimens, and then sought to develop a national project in integrated pest management under the umbrella of the International Biological Program. It would seem, therefore, reasonable to assume that they had come to share with systems ecologists a common ecological philosophy.

SYSTEMS ECOLOGY AND MODELING INSECT POPULATIONS IN AGRICULTURAL ECOSYSTEMS

Notwithstanding the rhetoric about the International Biological Program as a grand experiment to demonstrate the practical importance of more support for ecology, the National Science Foundation was quite reluctant to back the integrated pest management project. It was far too practically oriented for the Foundation, and, in fact, the director of the Foundation's Ecosystem Analysis Program, Jerry Franklin, indicated that his organization was not interested in the economic decision models, the "economic threshold" models, being developed by the scientists on the integrated pest management project. It was far more inclined to support the development of better models of the interactions between plants and herbivores, and to let the economic work be done by someone else.[40] There were, however, other reasons beside these institutional considerations for discontent at the National Science Foundation, the chief one being doubt about the relationship between integrated pest management and systems ecology.

According to the ecologists appointed by the National Science Foundation to review the development of the integrated pest management project and a proposal for its continuation into 1978, the project did not appear to be guided by the principles of systems ecology. They argued that many of the sub-projects were often organized into two separate and distinct phases: ecological research and formulation of models based on systems analysis. They objected to this division because, in their view, the unique feature of systems ecology was that mathematical modeling played an integral part in the process of discovering the structure of ecosystems. According to Watt, systems analysts and biologists began their investigation with no knowledge of ecosystem structure, only data. They developed an understanding of the structure as they analyzed the data. Analyses of variance would reveal which components of the system were more important, and then the biologist would investigate the properties of these more important components. The direction of biological research evolved in step with the analytical process. But, in the integrated pest management project, the biologists seemed to be gathering field data on the dynamics of insect populations without consulting with the modelers on the kind of data that should be collected. The latter seemed to intervene only to develop computer programs that would simulate these dynamics.[41]

A still graver objection to the development of the integrated pest management project stemmed from the appearance that the collection of field data was guided by the theories of population ecology rather than systems ecology. One reviewer commented that much of the work was on the ecology of key insect populations with little regard for either community or ecosystem ecology.[42] Jerry Olson, from the Oak Ridge National Laboratory and one of the major figures in the development of systems ecology, went even further by questioning the extent of the investigators' commitment to an ecosystem perspective; in his view, the project was not much more than a study in population ecology, albeit a far more exhaustive one, and on a far grander scale than usual.[43] In fact, the strongest supporter of the project was Fred Smith's colleague and intellectual ally at the University of Michigan, the theoretical population ecologist Larry Slobodkin.[44]

It seemed, then, to many observers that Huffaker and his colleagues were developing a research program inspired by classical theories of population ecology, and were then relying on the strategy of systems analysis to develop the economic decision models needed to develop effective integrated pest management programs. The relationship between ecological research and systems analysis must have been viewed quite differently by systems ecologists and the entomologists on the integrated pest management project. To account for this difficulty we must first turn to ideas of the most articulate critic of systems analysis in ecology.

CLASSIFYING ECOLOGICAL MODELS

Many theoretical population ecologists were very critical of the rapid expansion of support for systems ecology, all under the banner of improved environmental management. While conscious of the widespread accusations that ecologists were acting like passive bystanders as the world ran to its final environmental catastrophe, they claimed that this expansion was based on commitments that could not be met. Moreover, they feared that these commitments threatened to undermine the carefully nurtured establishment of ecology during the 1950s as a respectable academic discipline.[45] Nevertheless, by the end of the 1960s, systems ecologists enjoyed extraordinary levels of funding through an especially established division of the National Science Foundation, the Ecosystems Analysis Program, whose funding dwarfed that for the General Ecology Program.[46] This situation was especially aggravating for theoretical population ecologists, who saw Ken Watt and his colleague in the Canadian Department of Agriculture, Buzz Holling, trying to extend the domain of systems ecology to encompass their own specialty and carry with them such influential figures as Fred Smith and Larry Slobodkin.

From the mid-1960s onward, Richard Levins, a population biologist at Harvard University, became the principal, most forceful, and most unrelenting critic of systems ecology.[47] He and a small group of population biologists, including Richard Lewontin, Robert MacArthur, Ed Wilson, and Slobodkin himself, had been involved in efforts to develop a mathematically rigorous approach to the study of populational phenomena that was quite distinct from that of Watt and Holling by being fundamentally Nicholsonian, and also concerned with the evolutionary dimensions of these phenomena.[48] Levins also brought to bear on these efforts a Marxist perspective developing out of the political radicalism of the late 1960s.[49]

In 1968, Levins wrote an extremely critical review of Watt's *Ecology and Resource Management*.[50] The aim of Watt's book had been to illustrate how the field of resource management could benefit from the applications of systems ecology, chiefly to develop models to predict the impacts of alternative management strategies. Levins agreed with Watt that effective management of natural resources had to be built on a broader ecological perspective than usual. He also acknowledged that the interests and theories of population ecologists like himself had developed in directions not conducive to the development of general strategies for the management of these resources. However, he questioned Watt's claim that systems theory was uniquely suited to provide the required ecological perspective, and he also questioned the notion that systems ecology could ever provide a solid foundation for the future development of ecologically sound natural resource management strategies.

Levins first challenged the notion that systems ecology, as conceived by Watt and the proponents of International Biological Program, was in any sense more holistic than traditional population or community ecology. Although he admitted that systems ecologists appeared to be more aware of the systemic properties of ecosystems, he severely criticized the modeling strategy they adopted to describe these features. It seemed to him that systems ecologists merely reduced the ecosystem to its components, then reassembled the more important and quantifiable ones so that they might be able to write a workable computer program simulating the relationship between these components, as precisely as possible. This approach struck Levins as hardly holistic; in fact it appeared to him as a "peculiar empiricist, even reductionist, holism which tries to deal with complex wholes but without using any of the properties of complex systems".[51] Having dismissed systems ecologists' politically powerful claim to holism (organismic biology, during the 1960s and '70s, was marked by a strong rebellion against the reductionism of increasingly haegemonic molecular biology), Levins then turned to another supposed aspect of this school of thought. He and other critics of systems analysis in ecology drew a categorical distinction between "simulation" and "model". They argued that the function of a model was to provide a distillation of the essential features of complex ecological phenomena, and that the very failure of these models to match observed phonemena served to direct further investigations towards a sharper definition of the causal determinants of the same phenomena. It seemed to Levins that system ecologists' simulations could provide more accurate representations of the observed phenomena, but, by virtue of their very accuracy, they could not provide any direction for further analysis, and were, therefore, something like a dead end. Systems analysis was nothing more than a complex exercise in "curve fitting" to obtain precise numerical solutions to specific problems. Hence, it seemed to Levins that Watt was suggesting that the development of rational management strategies should be based simply on generalizations from case studies rather than on the development of sounder general ecological theory.[52] He believed that this attitude was bound to lead to a widening gap between ecologists and natural resource managers at a time when the environmental crisis required a closer alliance of the two groups.

Levins presented this criticism of systems analysis in ecology in a still more forceful, and more formal manner in his monograph, *Evolution in Changing Environments*.[53] He argued that ecological models could be characterized by their generality, realism, and precision. An ecological model was "general" when it was applicable to all species, regardless of their differences; it was "realistic" when all the assumptions underlying the model were deemed to be biologically sound; and finally, an ecological model was "precise" when the relationships between the different variables were numerically exact. Levins claimed that no model could exhibit all three features; at most, he argued, a model could meet two of the three. He classified the models of physicists turned theoretical population ecologists, such as Egbert Leigh at Columbia University, as general and precise.[54] They were therefore unrealistic and of little interest to ecologists. The models of ecologists such as Watt and Holling, he classified as precise and realistic.[55] Hence, Levins claimed, these models could not lead to any general understanding of ecological systems and were again of little interest to ecologists.[56] Levins obviously thought of his own general and realistic models as the superior ones. He was scarcely concerned with their imprecision, because he and other leading theoretical population ecologists, such as William Murdoch and John Connell, at the University of

California at Santa Barbara, felt that their discipline was far too young and immature to aspire to such methodological completeness.[57]

This classification of ecological models seemed to bring some order to the conflicting strategies adopted during the 1960s by mathematically inclined ecologists like Leigh, Watt, Holling, and Levins. All the uncertainty about the proper way to model ecological systems seemed to vanish by attributing the differences in modeling strategies to the different goals of the modelers: ecologists who sought to understand the most general aspects of population, community, and ecosystem structure, would have to adopt perforce the style preferred by Levins.[58] Those applied biologists who were only interested in predicting the behavior of a specific ecosystem for utilitarian purposes alone, could adopt the approach developed by Watt and Holling.

This classification also sheds some light on the controversial role of systems analysis in integrated pest management. The entomologists in Riverside and Berkeley, who counted among the leaders of the integrated pest management project, had often prefaced their discussions of integrated control programs by referring to John Nicholson's theory of population dynamics and had provided important experimental evidence in its favor. They seemed, therefore, to be far more sympathetic to theoretical population ecologists than to population ecologists turned systems ecologists such as Watt and Holling. On the other hand, the practical imperatives of integrated pest management forced an alliance between its proponents and system ecologists. The success of their work depended on the development of models of ecological systems that were sufficiently realistic and precise to allow relatively accurate predictions of population phenomena in specific agricultural ecosystems under specific control regimens. Without them, there could be no "economic thresholds". Since Carl Huffaker and his colleagues, especially Powers Messenger, acknowledged that physical environmental factors determined the capacity of the environment to support insect populations, they could not be excluded from their models.[59] These physical factors, however, were excluded from the models of theoretical population ecologists. Watt's and Holling's modeling technique, on the other hand, did not exclude them, and its application was, therefore, of much greater interest to Huffaker and the other scientists involved in the integrated pest management project.[60]

In sum, the modelers on Huffaker's team were simply transcribing those theories of population ecology that Watt and Holling had rejected as overly simplified representations of nature into a FORTRAN program, and then simulating the effects of various environmental manipulations to estimate their impact on the dynamics of specific key pests in specific agro-ecological contexts. This then helped to design optimal strategies for managing the use of chemical insecticides and other tactics against these pests. Systems analysis, so construed, served only to understand how a specific set of environmental circumstances affected the dynamics of particular populations; it was being used in the manner intended by industrial engineers who originally developed operations research. In other words, Huffaker's team was borrowing from Watt and Holling only their methods, not their philosophical commitments. This explains the perceptions of the reviewers from National Science Foundation that there was an unacceptable split between field research and theoretical development among these investigators. It also appears to confirm Levins' view that the adoption of systems theory to model ecosystems could only be motivated by utilitarian concerns.

DATA, THEORIES AND THE COMPLEXITY OF NATURE

By the 1960s, as we have seen, there were a number of schools of ecology which sought to replace an older tradition of natural history by discovering general ecological principles to explain how complex ecological communities are structured. Each of these school proposed its own modeling strategy to draw out the implications of these principles. Richard Levins sought to clarify the differences between these schools by developing a taxonomy of ecological models. His rather elegant taxonomic ordering suggested that systems ecologists were not interested in discovering any general principles. Thus, when he and William Murdoch were called upon to review the Carl Huffaker's integrated pest management project in preparation for the report to the National Academy of Science, *Pest Control: An Assessment of Present and Alternative Technologies*, they were quite critical of the ecological and modeling work being supported under its umbrella for its reliance on systems analysis.[61] Yet, there were good grounds for doubting the validity of Levins' argument against systems analysis. In principle, there was no reason why a very complicated world could not be conceived in terms of a model that was general, realistic and precise. To be sure, Levins' models tended to be more general and less realistic than Ken Watt's and Buzz Holling's, but this was simply a compromise imposed by the limited capabilities of the human mind, a situation which Watt and Holling sought to overcome with the aid of increasingly powerful computers. Theoretical population ecologists and systems ecologists were just as committed to the search for generality in nature as their critics.[62] By the late-1970s, after George Williams' damning refutation of holism in ecology and evolutionary biology, and the end of all talk (at least publicly) about wholes and irreducibility, the division between all these ecologists came to attributed to differences over ideas about the proper relationship between data and theory, between rationalism and empiricism.[63]

This view was forcefully articulated by Andy Gutierrez, one of the principal modelers for the integrated pest management project. Gutierrez learned about entomology and population ecology from Robert Van de Bosch and Carl Huffaker, as an undergraduate student in the department of entomology in

Riverside and as a graduate student in the department in Berkeley. He then went on to learn about ecological modeling from Holling, who by the mid-1960s had established a close relationship with the scientists in the Division of Biological Control at Berkeley. Gutierrez, still today a member of the embattled Division, admits that he sought to develop models that were realistic and precise because this was a requirement for the development of effective integrated pest management programs, and that systems analysis provided the most appropriate approach for this purpose. He also admits that his models were rather crude by the standards of "mathematicians" such as Levins. He questions, however, Levins' claim that the adoption of systems analysis to model ecosystems could only be motivated by utilitarian considerations, and attached far greater ecological significance to his models than Levins' classification would grant them.[64]

In *Ecological Relationships*, a textbook which he co-authored in 1976 with three other ecologists, and dedicated to Holling among others, Gutierrez summarized his position by focusing on Levins' three criteria and challenging the importance of Levins' own approach to ecology. He argued that the modeling strategy of theoretical population ecologists, which bore a close resemblance to that of mathematical physicists, had been unsuccessful in furthering understanding of the structure of ecological systems as they had rarely been substantiated by field data. He suggested rather sarcastically that these models described a very restricted number of ecological systems which existed only on paper.[65] Lastly, and most significantly, Gutierrez and the other authors of *Ecological Relationships* also suggested that nature was so much more complex than theoretical population ecologists supposed that, with the current dearth of detailed ecological information, it was simply inconceivable to even begin to think about generally valid ecological models. They advocated, therefore, the adoption of systems analysis, which would lead to the development of models that were perhaps less generally valid than those of theoretical population ecologists, but were far more easily tested against the "real world".[66]

This argument for greater integration of field data and ecological theory was summarily dismissed by the British theoretical population ecologist John Lawton, and by many other such ecologists beside Lawton.[67] For Lawton, *Ecological Relationships* was a manifesto for "FORTRAN ecology", a dismissive reference to the computer programming language used by its authors. He and his allies objected to any reliance on systems analysis and computer simulations to resolve the problems raised by Gutierrez and his colleagues. These methods had no place in ecology, except perhaps in "applied" ecology. For them, the final and true test of the validity of a model was not its empirical adequacy, but its "robustness".[68] Inspired by his Marxist outlook, Levins had argued in 1966 that a model was to be tested by examining its greater or lesser sensitivity to change of the underlying assumptions and its compatibility with models of altogether different ecological phenomena. Truth was reached through the dialectic of the different models; in other words, it was "the intersection of independent lies".[69] This strongly rationalist line of thinking was accompanied by such intense policing of population ecologists' publications against any more empiricist heresies that in 1974 the population ecologists Leigh Van Valen and Frank Pitelka complained on the pages of *Ecology* about a veritable reign of intellectual censorship.[70]

By the late 1970s, as the animosity and sharp divisions created by the International Biological Program and the subsequent patterns of funding began to dissipate, many ecologists whose concerns were far removed from the practical problems confronting Gutierrez began to pay more attention to the questions he and the other authors of *Ecological Relationships* had raised about the contemporary development of theroretical ecology. For example, the mathematical biologist from the University of California at Berkeley, George Oster, collaborated with Gutierrez to develop a more empirically adequate picture of predation than that produced by Levins and those theoretical population ecologists aligned with him.[71] Gutierrez' models of the interactions between plants, chiefly cotton plants in the Imperial Valley of California or in Nicaragua, and herbivores such as the Lygus Bug and Boll Weevil were also reviewed quite favorably by Donald Strong, a population ecologist then at the University of Florida.[72] Like Gutierrez, Strong and his colleague Daniel Simberloff had often decried the apparent trend toward de-coupling of theory from experimental testing as another deleterious effect of ecologists' aspirations to shape ecology after physics. At some point, they claimed, theories had to be tested against the systems they were supposed to represent, for otherwise nothing would restrain them from becoming meaningless mathematical formalisms. The separation of theoretical and experimental ecology was, according to Strong and Simberloff, nonsensical. Moreover, they believed that ecologists should prove that the world is not intrinsically idiosyncratic before striving to articulate general statements about the nature of ecological systems.[73] Ironically, Strong referred to Gutierrez' models during the symposium on integrated pest management and ecology organized by Marcos Kogan in San Antonio, to which Gutierrez himself was not invited.

MIGRATING FROM ENTOMOLOGY TO ECOLOGY

The population ecologist William Murdoch was asked, in 1988, about his views on the relationship between integrated pest management and ecology. He could speak with some authority about this subject since he and Richard Levins reviewed the organization and progress of the national integrated pest management project for the National Academy of Sciences's report *Pest Control: An Assessment of Present and Alternative Technologies*. He suggested that in some way the two ventures were quite distinct, stating that,

> the connection between integrated pest management and ecological theory is certainly not tight ... But there is certainly a connection between ecology and integrated pest management. Ecology is after all a way of asking questions about natural systems — figuring out what affects birth and death rates, using models to project what might

> happen if conditions, and the rates change; being aware that pest populations can compensate and if you spray them at one point they may survive better as a consequence at some other time; that if you spray for an early pest you may interfere also with predators of a later pest (causing upsets), so that you need to think in an integrated way about the potential interactions that can occur within the agricultural system (community) you are working with. This is, I think, the kind of connection that exists ... Ecology is not to pest management what physics is to engineering, and for good reason — the problems are much harder, the biological world much less predictable than the forces of gravity, and ecology a much younger science.[74]

In other words, according to Murdoch, the relationship between ecology and integrated pest management defied any definition because ecology lacked any coherent and unifying philosophy, a view which he had already defended in 1973. This response, in many ways, echoed Marcos Kogan's complaint that "ecological theory does not stand for integrated pest management as particle physics stands for the development of practical uses of nuclear energy'.[75] Both of them looked forward to the day when the relationship between ecology and integrated pest management would imitate that which they thought existed between particle physics and nuclear engineering. Consequently, that approach to ecological problems which most closely resembled physics should eventually play this leading role. And, in fact, in 1972, while discussing the role of ecology in developing more effective integrated pest management programs with the members of the Planning Committee preparing the report to the National Academy of Sciences on pest control, Murdoch had argued that,

> the ... problem ... may be best solved by forging much closer links between two disciplines, integrated control and population ecology, which have been separate until now ... university ecology centers could provide at short notice a good proportion of the manpower needed.[76]

He ignored, then, any possible contributions by systems ecologists, believing that the further development of population ecology was the sole prerequisite for the successful development of integrated pest management. Indeed sixteen years later, he had not changed his mind, still stressing that "systems analysis as applied ... in the [integrated pest management project] was not the best science, and work of that genre was not likely to be a lot of use".[77] Kogan, again, seemed to agree with this perspective as no ecologists or entomologists actually involved with the development of models for integrated pest management programs were invited to the conference in San Antonio on the relationship between ecology and integrated pest management.[78]

It seems, then, that the relationship between ecology and integrated pest management was problematic for Murdoch and Kogan because defining ecology was very difficult during the 1960s and '70s. There was no one commonly shared theory of ecology. Instead, new approaches to studying the interactions of insect populations in complex environments followed one another in rapid succession, each capturing different ecologists' imagination, and none displacing another. There was the approach proposed by the population ecologists turned systems ecologists Ken Watt and Buzz Holling, which was predicated on ideas about population dynamics that contradicted those of theoretical population ecologists such as Levins. They rejected these latter ecologists' ideas because they seemed to reflect so poorly the complex, organic world encountered by field biologists and agricultural scientists. Ecological theory, according to Watt and Holling, needed to be based on data rather than logical derivations from fundamental, and doubtful, presuppositions.

While the entomologists in the Division of Biological Control in Berkeley were sympathetic to these supposedly problematic presuppositions, especially as they were articulated by John Nicholson, Watt and Holling's modeling strategy seemed to open the possibility of transforming the simple models built upon them into meaningful economic models for integrated pest management programs. During the mid-1960s, the International Biological Program, the crowning attempt by of systems ecologists to transform the place of ecology in the political economy of American science, offered a unique opportunity for the institutional development of integrated pest management, by helping to cement a formal alliance between Watt, Holling and these entomologists in Berkeley. Theoretical population ecologists, many of whom had been extremely concerned to contribute to the resolution of one of the most visible environmental problems of the 1960s, had very little to offer to researchers in integrated pest management. The opportunities for institutional expansion which so concerned Murdoch were lost to systems ecologists. However, the mixing of seemingly incompatible theories of population ecology allowed by Watt and Holling's approach, and the grounding of this mixing in the practical problems of integrated pest management, offered theoretical population ecologists rich material to undermine the alliance. Levins' damning argument against Watt, Holling, and Gutierrez' empiricist critique of theoretical population ecology appealed to the widely held notion that researchers involved with the solution of practical problems cannot articulate coherent and theoretically significant ideas.

Watt, Holling, and Gutierrez were, of course, concerned with practical problems. The first two developed their ideas while in the service of the Canadian Department of Agriculture, working to develop better ways to control the Spruce Budworm and other insects damaging Canadian forests, and then became involved in more general problems of environmental management and urban planning. Gutierrez was, and still is, employed by the Division of Biological Control. He has admitted that his own empiricism was in part motivated by the need to articulate models of greater predictive power than those of most theoretical population ecologists. Yet, the acceptance during the late 1970s of the critiques of theoretical population ecology articulated by Watt, Holling, and Gutierrez among population ecologists with very little direct interest in practical applications of ecological models, suggests that there was not the necessary connection between their ideas and their institutional location suspected by Levins. Watt, Holling, and Gutierrez were ecologists just as much as Levins and their other critics, the only difference being that they were associated at one time or another with institutions dedicated to the solution of practical, agricultural problems. On the other hand, as Levins has pointed out, Gutierrez and others' "distinction between systems theory as a philosophy and as a tactic holds up formally, but as historical fact the practice easily generalizes into a latent philosophy".[79] For Gutierrez, trained as he was during they heady days of the International Biological Program, and Carl Huffaker's opportunistic exploitation of the latter to create a national integrated pest management program, the integration of the Nicholsonian ecology, which had bolstered the institutional stability of biological control, and the strategies to model the more complex world seen by Watt, Holling, and systems ecologists generally, became a fundamental philosophical commitment. It was not simply an accommodation to fulfill the tradition of service which had characterized so much of Harry Smith's work in the Division of Biological Control, but a commitment to a new mode of inquiry into important ecological issues, such as the relationship between herbivores and plants. To such classic, and ever contentious, questions such as the one asked by Nelson Hairston, Fred Smith, and Larry Slobodkin in 1960, "why is the world green?", Gutierrez answered that ecologists had to draw on both population and physiological ecology — chiefly the effects of the physical environment on the reproductive behavior of insects. As Kogan himself pointed out, "integrated pest management often advances ahead of the establishment and, at times, despite the contradictions, of current ecological theories".[80] Gutierrez was an ecologist who saw no critical problem in the relationship betwen integrated pest management and ecology. Of course, such a situation was quite worrying to those like Kogan and Murdoch who believed instead that theoretical population ecology should stand to integrated pest management as they supposed particle physics stood for "engineering" or "the practical applications of nuclear energy".[81] Until that day arrived, they would always complain that "integrated pest management lacks a solid theoretical background".[82] Needless to say, this was no longer a debate about how to better control insect pests, but about how to best model nature: Gutierrez had moved a long way from the world of his "grandfather", Harry Smith. For others, these entomologists and ecologists' preoccupations should stimulate a more critical consideration about the variety of factors shaping the ever-changing relationships between science and technology more generally.

NOTES

1 R.L. Rabb, "Introduction to the conference" in Rabb & F.E. Guthrie (eds), *Concepts of Pest Management* (Raleigh: North Carolina State University, 1971), pp 1–5, on p. 3.

2 M. Kogan, "Preface" in Kogan (ed.), *Ecological Theory and Integrated Pest Management Practice* (New York: Wiley, 1986), pp. ix–xi, on pp. ix–x.

3 For a rather critical appreciation of the status of American ecology as a scientific discipline, see C.H. Waddington, "The origin" in E.B. Worthington, *The Evolution of IBP* (Cambridge: Cambridge University Press, 1975), pp. 4–11, on p. 9. See also J.B. Hagen, *An Entangled Bank: The Origins of Systems Ecology* (New Brunswick: Rutgers University Press, 1992).

4 E.P. Odum, "The new ecology", *BioScience* 14 (1964): 14–6.

5 Peter Taylor has focused his attention on the metaphors adopted by Howard Odum, and his work raises important questions about defining Odum's ecological philosophy as either "organicist" or "mechanistic"; see "Technocratic optimism, H.T. Odum and the partial transformation of ecological metaphor after World War 2", *Journal of the History of Biology* 21 (1988): 219–44.

6 Hagen, *Entangled Bank*, pp. 130–6.

7 E.P. Odum, *Fundamentals of Ecology* (Philadelphia: W.B. Saunders, 1959), p. 11. In the discussion of Eugene Odum's ideas I will refer to this second edition of the textbook since his philosophical perspective is presented here much more forcefully than in the first one (1954).

8 Odum, *Fundamentals*, pp. 7 and 25–6.

9 Hagen, *Entangled Bank*, pp. 158–62.

10 H.T. Odum, *Environment, Power, and Society* (New York: Wiley, 1971), p. 60.

11 Odum, "New Ecology". See also Odum, *Environment*, pp. 254–73.

12 See S.E. Kingsland, *Modeling Nature. Episodes in the History of Population Ecology* (Chicago: University of Chicago Press, 1985).

13 T. Dobzhansky, "Comments on the symposium", *Cold Spring Harbor Symposia on Quantitative Biology* 22 (1957): 235; see also the editorial footnote to the article by A. Milne, "Weather, enemies and natural control of insect populations", *Journal of Economic Entomology* 52 (1959): 532–3.

14 H.G. Andrewartha, "The use of conceptual models in population ecology", *Cold Spring Harbor Symposia on Quantitative Biology* 22 (1957): 219–32, on p. 231; see also, in the same volume, on pp. 203–15 and 253–67, L.C. Birch, "The role of weather in determining the distribution and abundance of animals" and A. Milne, "Theories of natural control of insect populations".

15 K.E.F. Watt, "Studies in population productivity: I. Three approaches to the optimum yield problem in populations of *Tribolium confusum*", *Ecological Monographs* 25 (1955): 269–90.

16 On the history of research in population ecology in the Fisheries Research Board, see W.E. Ricker, "The Fisheries Research Board of Canada: seventy-five years of achievements", *Journal of the Fisheries Research Board of Canada* 32 (1975): 1465–90, on pp. 1473–7.

17 K.E.F. Watt, "Mathematical population models for five agricultural crop pests" in E.J. Le Roux (ed.), "Agricultural and forest insect dynamics", *Memoirs of the Entomological Society of Canada* 32 (1963): 1–103, on pp. 83–91.

18 K.E.F. Watt, "The systems point of view in pest management" in Rabb & Guthrie, *Concepts of Pest Management*, pp. 21–34.

19 See K.E.F. Watt, "Use of a computer to evaluate alternative insecticidal programs", *Science* 133 (1961): 706–7.

20 K.E.F. Watt (University of California), interview with Paolo Palladino, 24 July 1991.

21 For a history of operations research, see M. Fortun & S.S. Schweber, "Scientists and the legacy of world war II: the case of operations research (OR)", *Social Studies of Science* 23 (1993): 595–642.

22 K.E.F. Watt, "Use of mathematics in population ecology", *Annual Review of Entomology* 7 (1962): 243–60.

23 K.E.F. Watt, "The nature of systems analysis" in Watt (ed.), *Systems Analysis in Ecology* (New York: Academic Press, 1966), pp. 1–14 on pp. 1–3.

24 See K.E.F. Watt, "Ecology in the future" in Watt, *Systems Analysis in Ecology*, pp. 253–67.

25 Watt's holism may be traced to his doctoral training at the University of Chicago, where he was Thomas Park's student. As Greg Mitman has shown, between the 1930s and '50s, Park and other associated ecologists and sociologists developed a school of thought about animal and human communities that was very strongly holistic; see *The State of Nature: Ecology, Community, and American Social Thought, 1900–1950* (Chicago: University of Chicago Press, 1992).

26 R.P. McIntosh, *The Background to Ecology. Concept and Theory* (Cambridge: Cambridge University Press, 1985), p. 228.

27 Archives of the University of Georgia: Ecological Society of America: Institute of Ecology: E.P. Odum Files: Watt to Odum, 11 January 1965.

28 V.M. Stern *et al.*, "The integrated control concept", *Hilgardia* 29 (1959): 81–101, on p. 86.

29 Stern *et al.*, "Integrated control", p. 94.

30 R. Van den Bosch & V.M. Stern, "The integration of chemical and biological control of insect pests", *Annual Review of Entomology* 7 (1962): 367–386.

31 See D.A. Chant, "Integrated control systems" in C.E. Palm (ed.), *Scientific Aspects of Pest Control* (Washington: National Academy of Sciences, 1966), pp. 193–218, on pp. 207 and 210.

32 See O.L. Loucks, "The United States International Biological Program: An ecosystem perspective after fifteen years" in N. Polunin (ed.), *Ecosystem Theory and Application* (New York: Wiley, 1986), pp. 390–405; W.F. Blair, *Big Biology: The US/IBP* (Stroudsburg: Dowden, Hutchison and Ross, 1977), pp. 162–165.

33 See N.G. Hairston, F.E. Smith & L.B. Slobodkin, "Community structure, population control, and competition", *American Naturalist* 94 (1960): 421–5; and L.B. Slobodkin, F.E. Smith & N.G. Hairston, "Regulation in terrestrial ecosystems and the implied balance of nature", *American Naturalist* 101 (1967): 109–24.

34 F.E. Smith, "The IBP and the science of ecology" in US House of Representatives, *International Biological Program*, Hearings before the Subcommittee on Science, Research and Development, Committee on Science and Astronautics, 90th Congress, First Session, (Washington: United States Government Printing Office, 1967). In *Big Biology*, Frank Blair presents an account of the development of International Biological Program in the United States. For a more comprehensive review of the program, systems ecology, and the controversies surrounding them, see McIntosh, *Ecology*, pp. 193–241. For a historian's view on the subject see C. Kwa, "Representations of nature mediating between ecology and science policy: The case of the International Biological Programme", *Social Studies of Science* 17 (1987): 413–442.

35 United States National Committee for the International Biological Program, *United States Participation in the International Biological Program* (Washington: National Academy of Sciences, 1974), p. 19.

36 Kwa, "Representations", p. 415.

37 D. Simberloff, "A succession of paradigms in ecology: essentialism to materialism and probabilism" in E. Saarinen (ed.), *Conceptual Issues in Ecology* (Dordrecht: Reidel, 1982), pp. 63–99, on p. 88.

38 Dale Bottrell (University of Maryland), interview with Paolo Palladino, 20 September 1988.

39 For an assessment of the role of systems analysis in pest control during the early phases of the development of integrated control and integrated pest management, see R.W. Stark & R.F. Smith, "Systems analysis in pest management" in C.B. Huffaker (ed.), *Biological Control* (New York: Plenum Press, 1971), pp. 331–45.

40 Files of W. Klassen (henceforth Klassen Files): NSF-IBP/RANN-Dr. Huffaker: C.H. Hoffman to H.O. Graumann, 26 September 1973, p. 11.

41 Klassen Files: NSF-IBP/RANN-Dr. Huffaker: Memorandum by W.E. Hazen, 3 April 1973.

42 Archives of the National Science Foundation: Proposal BMS 2504223: Reviewer's comments: Cotton sub-project", (ca. September 1974); and also J.T. Callahan to C.B. Huffaker, 20 December 1973.

43 Klassen Files: NSF-IBP/RANN-Dr. Huffaker: Letter from ENT/PSR Committee to T.Y. Edminster, 27 September 1971; and J. Olson, "Review of National Science Foundation proposal P2B1254–000", 31 January 1972. Also Stanley Auerbach (Oak Ridge National Laboratory) to Paolo Palladino, 21 October 1988.

44 Carl Huffaker (University of California), interviews with Paolo Palladino, 15 and 18 August 1988.

45 See McIntosh, *Ecology*, pp. 308–23.

46 Hagen, *Entangled Bank*, pp. 178–81.

47 R. Levins, "The strategy of model building in population ecology", *American Scientist* 54 (1966): 421–431.

48 See U. Segerstrale, “Colleagues in conflict: an in vivo analysis of the sociobiology controversy”, *Biology and Philosophy* 1 (1986): 53–87.

49 For a review of the literature on Marxism and American radicalism in the 1960s, see T. Wohlforth, “The Sixties in America:, *New Left Review* 178 (1989): 105–123. It is unfortunate that to this day there has been no systematic study of the emergence during the 1960s of an American Marxist scientific tradition centered around the School of Marxist Education in New York, except for D.L. Hull, *Science as Process* (Chicago: University of Chicago Press, 1988), pp. 223–30.

50 R. Levins, “Ecological engineering: theory and technology”, *Quarterly Review of Biology*, 43 (1968): 301–5. Levins’ later discussion of electric circuit analysis is also particularly relevant to the subject of systems ecology and holsim because Howard Odum had defended the uniqueness of ecosystems by relying on analogies to just such circuits; see R. Levins, “Complex systems” in C.H. Waddington (ed.), *Towards a Theoretical Biology* (Edinburgh: Edinburgh University Press, 1970), Vol. III, pp. 73–88.

51 Levins, “Ecological engineering”, p. 303. For a further elaboration of this line of argument see R. Levins & R.C. Lewontin, *The Dialectical Biologist* (Cambridge: Harvard University Press, 1985), pp. 132–60; and for Watt’s later thoughts on systems analysis in ecology, see K.E.F. Watt, “Critique and comparison of biome ecosystem modeling” in B.C. Patten (ed.), *Systems Analysis and Simulation Ecology* (New York: Academic Press, 1972), Vol. III, pp. 139–52.

52 Levins, “Ecological engineering”, p. 302.

53 R. Levins, *Evolution in Changing Environments* (Princeton: Princeton University Press, 1968), pp. 3–9.

54 See E. Leigh, “On the relation between productivity, biomass, diversity and stability of a community”, *Proceedings of the National Academy of Sciences* 53 (1965): 777–83.

55 See K.E.F. Watt, “The choice and solution of mathematical models for predicting and maximizing the yield of a fishery”, *Journal of the Fishery Research Board of Canada* 13 (1956): 613–45; and C.S. Holling, “The components of predation as revealed by a study of small-mammal predation of the European Pine Sawfly”, *Canadian Entomologist* 91 (1959): 293–320.

56 Levins, “Complex systems”.

57 See Levins, “Ecological engineering”, and W.W. Murdoch & J. Connell, “All about ecology” in I.G. Barbour (ed.), *Western Man and Environmental Ethics* (Reading: Addison Wesley, 1973), pp. 156–70; for a much harsher appreciation of systems analysis in ecology, see R.M. May, *Stability and Complexity in Model Ecosystems* (Princeton: Princeton University Press, 1973), pp. 3–12.

58 For alternative views on the classification of ecological models, see C.A.S. Hall & J.W. Day, “Systems and models: terms and basic principles” in C.A.S. Hall & J.W. Day (eds), *Ecosystem Modeling in Theory and Practice*

(New York: Wiley, 1977), pp. 5–36; C.S. Holling, "Resilience and stability of ecological systems", *Annual Review of Ecology and Systematics* 4 (1973): 1–23; and C.J. Walters, "Systems ecology: the systems approach and mathematical models in ecology" in E.P. Odum, *Fundamentals of Ecology* (Philadelphia: Saunders, 1971), pp. 276–92.

59 See P.S. Messenger, "Bioclimatic inputs to biological control and pest management programs" in Rabb & Guthrie, *Concepts of Pest Management*, pp. 84–102.

60 See E.W. Waters & B. Ewing, "Development and role of predictive modeling in pest management systems: insects", C. Shoemaker, "Management models for integrated pest control: Mathematical structure and solution", and C.B. Huffaker *et al.*, "The need for systems analysis and its use in the US/IBP integrated pest management project", all in R. Tummala *et al.* (eds), *Modeling for Pest Management* (East Lansing: Michigan State University, 1976), pp. 19–27, 32–39, and 209–216.

61 Archives of the National Academy of Sciences: Environmental Studies Board: Studies: Problems of Pest Control: (A: C&B: ESB: X: 1971–1973): Committee Studies: Corn/Soybeans Study Team: Planning Committee: W. Murdoch, 26 January 1972; and General: R. Levins to H. Engstrom, 21 November 1973.

62 See C.A.S. Hall & D.L. De Angelis, "Models in ecology: Paradigms lost or paradigms found?" *Bulletin of the Ecological Society of America* 66 (1985): 339–46; and also R.H. Peters, "From natural history to ecology", *Perspectives in Biology and Medicine* 23 (1980): 191–203.

63 See Hagen, *Entangled Bank*, pp. 151–63.

64 Andy Gutierrez (University of California), interview with Paolo Palladino, 19 August 1988.

65 N. Gilbert *et al.*, *Ecological Relationships* (Reading: Freeman and company, 1976), pp. 3–11 and 70–91.

66 Gilbert *et al.*, *Ecological Relationships*, p. 73.

67 J.H. Lawton, "Spokes missing in the ecological wheel", *Nature* 265 (1977): 768. The motives for Lawton's critical attitude are clearly presented in G. Caughley & J.H. Lawton, "Plant-herbivore systems" in R.M. May (ed.), *Theoretical Ecology. Principles and Applications* (Sunderland: Sinauer Publishers, 1981), pp. 132–166.

68 See W.C. Wimsatt, "False models as means to truer theories" in M.H. Nitecki & A. Hoffman (eds), *Neutral Models in Biology* (Oxford: Oxford University Press, 1987), pp. 25–55, as well as May, *Stability and Complexity in Model Ecosystems*; and Maynard Smith, *Models in Ecology* (Cambridge: Cambridge University Press, 1974).

69 Richard Levins, quoted in Wimsatt "False models as a means to true theories", p. 52.

70 L. Van Valen & F. Pitelka, "Commentary — Intellectual censorship in ecology", *Ecology* 55 (1974): 925–926.

71 See G. Oster, "Predicting populations", *American Zoologist* 21 (1981): 831–44; and H.Y. Wang, A.P. Gutierrez, G. Oster & R. Daxl, "A population model for cotton growth and development: coupling cotton-herbivore interactions", *Canadian Entomologist* 109 (1977): 1359–74.

72 D.R. Strong, "Population theory and understanding pest outbreaks" in Kogan, *Ecological Theory*, pp. 37–58.

73 See D.R. Strong, "Natural variability and the manifold mechanisms of ecological communities" in G.W. Salt (ed.), *Ecology and Evolutionary Biology* (Chicago: University of Chicago Press, 1984), pp. 56–80.

74 William Murdoch (University of California) to Paolo Palladino, 17 May 1988.

75 Kogan, "Preface", p. ix.

76 Archives of the National Academy of Sciences: Environmental Studies Board: Studies: Problems of Pest Control (A: C&B: ESB: X: 1971–1973): Planning Committee: W. Murdoch, 26 January 1972.

77 William Murdoch (University of California) to Paolo Palladino, 17 May 1988.

78 Andy Gutierrez (University of California), interview with Paolo Palladino, 19 February 1992.

79 Richard Levins (Harvard University) to Paolo Palladino, 17 May 1988.

80 Kogan, "Preface", p. ix.

81 Archives of the National Academy of Sciences: Environmental Studies Board: Studies: Problems of Pest Control (A: C&B: ESB: X: 1971–1973): Planning Committee: W. Murdoch, 26 January 1972; and Kogan, "Preface", p. ix.

82 Kogan, "Preface", p. ix.

CONCLUSION

PROTECTING CROPS AND THE ECOLOGY OF INSECT POPULATIONS

Before the 1960s, before the publication of Rachel Carson's enormously successful book, *Silent Spring*, problems of crop protection were, in North America, matters concerning mostly agricultural scientists in universities, experiment stations, and industrial laboratories. After *Silent Spring*, these problems became subjects of intense debate even among people far removed geographically, socialy, and institutionaly from the world of agriculture. The urbane president of the United States, John Kennedy, for example, learned about these problems while reading the equally urbane *New Yorker*. This debate did much to project ecology into the public limelight, rather ironically, as the new scientific savior in an age of increasing doubt about the beneficence of scientific, technological, and industrial progress. In 1965, in *Restoring the Quality of our Environment*, a report of the President's Science Advisory Committee to Kennedy's successor, Lyndon Johnson, the entomologists and ecologists David Pimentel, Donald Chant, Arthur Kelman, Robert Metcalf, Dale Newsom, and Carroll Smith questioned the previously largely unchallenged entomological dictum that "the only good bug is a dead bug". Ecology taught quite otherwise, and, thus, these scientists argued that the development of this science was central to the development of an environmentally sound approach to crop protection. They wrote more specifically that,

> In the *Origin of Species* Darwin stated that man does not know how the numbers of a single species are controlled in nature. A century later his statement needs little modification. The population biologist or ecologist interested in pest population control has an opportunity to contribute basic knowledge in population dynamics. Many of the concepts and principles of populations behavior and interspecific interaction arose from attempts to control some insect pest. Conversely, scientific study of behavior and interaction will contribute to applied problems of pest control.[1]

Since these entomologists and ecologists felt that economic entomology should be about reducing the numbers of insect or weed populations to levels where they no longer caused economic losses, without necessarily turning to ecologically destructive insecticides such as DDT, the most immediate question that they and economic entomologists should address was what has been, and perhaps still is, one of the most controversial questions in ecology, namely "how is the abundance and distribution of animal populations determined?"

Ecologists and ecologically inclined entomologists have provided many different answers to this question, but none of these answers has ever satisfied everyone. This situation has provided the ecologist Robert McIntosh with much evidence for the claim, in his exhaustively detailed history of ecology, that the latter is an intellectually fragmented discipline. However, my account of this fragmentation suggests that if it is indeed possible to build a more coherent discipline, this cannot be done as McIntosh hopes, that is, by familiarizing ecologists with just the intellectual history of their field.[2] Some of these ecologists, especially those among them in institutions dedicated to economic entomology, have long realized that many of the different answers to the central problematic of ecology have not been determined by the facts of "population behavior and interspecific interaction", but by other considerations closely tied to the respondents' particular institutional locations. Of course, as the sociologists Nigel Gilbert and Michael Mulkay would point out, these institutional explanations were seen to apply only to rival ecologists and economic entomologists' answers.[3] I have sought to further these views about the institutional structuring of truths about the organization of nature, or rather about the appropriateness of particular methods for understanding nature, but in a more symmetric manner. That is, I have considered in some detail how different groups of economic entomologists and ecologists have sought from the beginning of this century to answer the problem highlighted in *Preserving the Quality of our Environment*, and why these economic entomologists and ecologists ever cared about each other's answers, without, as they tended to, privileging one group's views over another's.

The first step was to explain the apparent contradiction between that formative role played by economic entomologists in the early evolution of ecological ideas, highlighted in *Preserving the Quality of our Environment*, and the gross ignorance of ecology which led them to promote the use of destructive insecticides like DDT so vividly and powerfully described in *Silent Spring*. Many historians and other observers before me have addressed this problem. The general consensus is that the history of these insecticides and the role of the entomological profession in their development is nothing but a chapter in the history of the transformation of American agriculture into a technologically advanced, capitalist enterprise: agricultural businesses and the chemical industry sacrificed farmers' economic welfare and everyone's natural environment for the sake of corporate power. Although I generally agree with this assessment, it is a reductive oversimplification of this history. The entomological profession was undoubtedly encouraged by the chemical industry to work on, and press upon farmers and other agricultural producers, the use of chemical insecticides. Yet, as the historians Conner Sorensen and John Perkins have suggested, more or less indirectly, the success of insecticides also rested on farmers' and entomologists' shared beliefs in technological progress and the control of nature.[4] These beliefs, in turn, can be viewed as important elements mediating the capitalist transformation of agriculture. There were, of course, many other such mediating factors. I have sought to recover some of them by examining, rather ironically, the careers and work of two communities of entomologists who shared a strong opposition to the prevailing philosophy that "the only good bug is a dead bug". One was in the state agricultural experiment stations at Riverside and Berkeley, and the other was dispersed across several federal research institutes of the Canadian Department of Agriculture. As the historians Malcolm Nicolson and Jonathan Harwood have argued in their work on the history of ecology and genetics, such a comparative approach is most useful to tease apart the variety of factors that shape scientific discourse, and the networks of responsibility for the insecticide crisis as well.[5]

ENTOMOLOGY IN THE UNITED STATES: FROM BIOLOGY TO CHEMISTRY

During the late nineteenth and early twentieth centuries, many American students of natural history established a strong professional identity as entomologists in the context of the Progressive effort to transform American agriculture into an industry as efficient as John D. Rockefeller's businesses.[6] Between the 1880s and the 1920s, many of these entomologists found employment in ever-expanding institutions for agricultural education and research. The justification for the public support they received in the land grant universities and agricultural experiment stations was to develop more effective methods of crop protection. This work was, and to a large extent still is, the mainstay of American entomology. The development of ever more powerful insecticides, and the gradual abandonment of other environmental manipulations, such as the introduction of insects preying on, or parasitizing, agricultural pests, or the modification of cultural practices, came in response to perceived and expressed desires of the entomologists' agricultural constituencies. This was not simply a case of entomologists bowing to farmers' preferences. By the early years of this century, land grant colleges across America were making farmers' sons and daughters into graduates eager to serve the very agricultural community from which most of them hailed. It clearly made sense to the entomologists among these graduates that the use of chemical insecticides was the most sensible way to proceed. Of course, the chemical industry stood to gain from farmers' negative attitudes toward the very constraining and labor intensive environmental manipulations proposed by some entomologists, and it encouraged them and entomologists in all manners possible to increase their use of the insecticides it manufactured. This included support for entomological research directed toward the development and improvement of insecticides. The entomologist William Morton Wheeler, at Harvard University, often argued against this trend by suggesting that,

> the economic entomologist, like the physician, should strive to make himself unnecessary by prophylactic measures, or perhaps I should say that he should prevent plant diseases rather than spend his time studying symptoms and applying remedies to diseases after they become established.[7]

Just as in medicine, however, Wheeler's advice failed to win any supporters. Thus, while botanists and zoologists, in institutional circumstances similar those of entomologists, became breeders and range or forest managers, and, according to Deborah Fitzgerald and Ronald Tobey, subsequently developed a strong professional identity as "biologists", more specifically, either as "geneticists" or "ecologists", entomologists themselves became "chemists".[8] This situation was a source of some embarrassment to some of their number in leading American universities such as Harvard and Yale, especially when they discovered that colleagues abroad seemed to approach questions of economic entomology in a quite different manner.

ENTOMOLOGY IN CANADA: ECONOMIC ENTOMOLOGY AS BIOLOGY OF INSECTS ON AGRICULTURAL CROPS

The early development of entomology in Canada was not significantly different from that of entomology in the United States. Students of natural history found employment as agricultural scientists, and developed a professional identity around their agricultural mission. Yet, Canada was, and in many ways still is, a very different nation from the United States. Notwithstanding the perennial tensions between the federal government and its provincial counterparts, the former clearly enjoyed the upper hand during the first half of this century. Furthermore, the Canadian parliamentary system isolated the central government and federal bureaucracy from local constituencies to a far greater extent than was the case in the United States. This situation, as the historian Marilyn Dubasak has argued, had profound implications for the development of Canadian policies to manage natural resources and protect wildlife.[9] Local constituencies faced almost insurmountable difficulties in their attempts to challenge any developers who enjoyed the support of central government. Grassroots campaigns, such as those in the United States, were virtually nonexistent. This political situation had similar, profound implications for the development of Canadian entomological research.

Just as in the United States, the vast majority of Canadian entomologists were employed in centers for agricultural research. However, the model for the development of agricultural research in Canada does not seem to have been an American one, but a hybrid between the latter and one imported from the British imperial metropolis.[10] The centers where the entomologists found their employment were under the centralized control of the Dominion Department of Agriculture. Agricultural extension and education were provincial responsibilities quite separate from research. Entomologists in the federal research centers were, therefore, far more removed from any significant contact with the agricultural community, and from the daily problems of operational development, than their American counterparts. Moreover, as with most industries in Canada, manufacturers of chemical insecticides tended to be subsidiaries of American ones. These manufacturers had, therefore, very little interest in supporting any entomological research in Canada; their needs were met quite satisfactorily by the work of entomologists in the United States. The consequence of these two situations was that it was relatively easy for federal entomologists to transform their brief from operational work to study of the supposed more abstract general foundations of this operational work. This new work was quite different from that of either their few counterparts in the provinces, or from that of the vast majority of American entomologists. While the latter two groups were more or less committed to the development and use of improved chemical insecticides to protect farmers' crops, entomologists in the research institutes of the Department of Agriculture could spend considerably more time studying the biology of insect species. Needless to say, the relative lack of channels of communication at the grass-root level was translated into a rapid divergence of interests between farmers and entomologists. The Dominion Entomologist Gordon Hewitt, was quite free to dismiss any questions about the utility of his ecological investigations, by threatening that,

> We [federal entomologists] are not here to listen to remarks of that description and we will withdraw immediately from this investigation unless you accept what we tell you as the truth.[11]

Farmers were equally free blissfully to ignore the work of Hewitt's federal entomologists. Canadian entomologists enjoyed far less support among these potential political supporters than did American entomologists in their own country. Departments of entomology in the newer universities across the Canadian western provinces were few, and courses in economic entomology were just as rare. Graduates in zoology interested in insects biology had to seek more advanced training in the United States (and, to some extent, in Great Britain), and, upon graduation, they found many more employment opportunities across the border than in Canada. Consequently, the Canadian entomological community did not grow as numerous as the American one. On the other hand, for those entomologists who remained in Canada, or returned there after earning their advanced degrees abroad, entomology was a sub-field of biology, rather than chemistry as was the case in the United States. One important feature of this alignment was that these entomologists were so far removed from the world of farmers that they could do very little to halt, if so inclined, the latter's growing dependence on the insecticides promoted by the chemical industry. DDT caused as much environmental damage in Canada as it did in the United States.

BIOLOGICAL CONTROL IN THE UNITED STATES AND CANADA

These different institutional situations of entomological research in the United States and Canada had a very definite impact on the evolution of one of the most promising alternatives to chemical control. Biological control, the purposeful introduction of predators and parasites to control insect pests, could in many cases be a powerful and extremely economical solution to farmers' problems. The citrus industry of southern California realized this very early on in the century, and, during the next fifty years entomologists in the relatively well supported Division of Biological Control in the Citrus Experiment Station at Riverside gained a considerable international reputation for its work in biological control. It was so successful that its program even managed to expand during the 1940s and '50s, a time when programs in biological control were virtually eliminated throughout the United States because it seemed so obvious that the new generation of synthetic insecticides that followed the development of DDT were the final answer to all problems of economic entomology. The exceptional endurance of biological control in California was the result of Harry Smith's careful nurturing of a close relationship between his Division of Biological Control and the citrus industry. However interested in problems of population ecology, and in all that work in biological control had revealed about these problems, Smith never let anyone in the Division forget that "the goals of the Division of ... Biological Control ... and the Experiment Station ... were to solve pest problems".[12]

Work in biological control also thrived in Canada, especially in the Dominion Laboratory for Biological Control at Belleville. Just as in the United States, by the early decades of this century biological control was no longer farmers' preferred approach to controlling insect pests. Chemical insecticides seemed far more attractive. However, those species damaging the Canadian forests, and the very precious lumber extracted therefrom, could not be controlled with such insecticides, because the cost of such an approach was prohibitive. The lumber and pulp industry was, therefore, supportive of work on alternative approaches. Yet, the rewards of this work were few, as there was no conspicuous success, comparable to that of the Vedalia Beetle against the Cottony Cushion Scale in the citrus groves of southern California. Biological control did not become a major item on the list of most Canadian entomologists' responsibilities. On the other hand, neither did the development of better insecticides become such a major item. Thus, when the new synthetic insecticides appeared on the market, work in biological control did not suffer the same fate it did in the United States. Research in biological control continued to expand, especially as it became a subject of considerable scientific interest when the shortcomings of DDT and other allied synthetic insecticides became inescapably obvious to most entomologists. During the 1950s and '60s, entomologists and ecologists in the Division of Forest Insects and the newly renamed Entomology Research Institute for Biological Control began to examine the ecological foundations of all strategies of crop protection, chemical, biological, and otherwise.

During the late 1950s, Smith's heirs in the Division of Biological Control at Riverside and Berkeley, and a group of entomologists at the Annapolis Royal Laboratory of the Canadian Department of Agriculture, led by Allison Pickett, sought to develop a novel approach to crop protection which would avoid the serious problems that were then arising from an ecologically indiscriminate use of chemical insecticides in both the United States and Canada. They proposed to use chemical insecticides in ways that would avoid any irreparable damage to those biological mechanisms regulating the abundance of insect populations, chiefly predators and parasites. Reflecting the orientation of Canadian entomological work, between 1946 and 1953, Pickett's group produced a number of seminal papers on the ecology of insects in apple orchards, but was unable to do much about the economic problems supposedly motivating its work. The entomologists at Riverside and Berkeley were unable to do much more but recommend the use of more selective chemical insecticides. Nevertheless, in 1959, four of their number, Vernon Stern, Ken Hagen, Ray Smith and Robert Van den Bosch, joined together to write a seminal paper on the "integrated control concept". This concept was widely viewed as a revolutionary innovation in the work of crop protection because it was grounded in the latest advances in ecology.

During the early 1960s, entomologists more or less closely associated with Belleville, chiefly Donald Chant, Albert Turnbull, Ken Watt and Buzz Holling, challenged the notion that work in biological or integrated control was at all grounded in any knowledge of the ecology of insect species. They mounted a sharp critique of the relationship between then current practices of biological control and ecological theory, focusing especially on the work of the leading entomologists in this field, those at Riverside and Berkeley. In contrast to them, the Canadian entomologists believed that the ability of predators and parasites to control insect populations was rather limited; Chant, in particular, went on to argue that chemical insecticides would inevitably have to be the mainstay of any integrated control program. They suggested more or less publicly that the supposedly questionable relationship existing between ecology and the work in biological control by the entomologists in Riverside and Berkeley was the result of institutional pressures, chiefly the need to demonstrate the economic viability of biological control at a time when it was coming under the severe scrutiny of those entomologists demanding greater support for work on chemical insecticides.

There is good evidence to suggest that the Canadian entomologists' claims about the connection between the institutional circumstances of the entomologists in Riverside and Berkeley and the latter's attitude toward the relationship between scientific theory and technological practice were not ill-founded. In fact, responding to their critics' recommendation of more extensive ecological investigation prior to initiating any operational work, the entomologists in Riverside and Berkeley, led by Carl Huffaker and Paul De Bach, wrote that,

> the time (many years) and money spent to make such exhaustive studies before making any introductions would be imprudent.[13]

It is, however, also evident that the Canadian entomologists' approach was very much shaped by their own institutional context. They were expected to adopt a theoretically sophisticated viewpoint on questions of technological development, because it was widely assumed in this institutional environment that technological development was crucially dependent on the direct application of an appropriate body of theoretical precepts. Applied scientists, for them, were supposed, by definition, to be guided by the best theory and were not supposed to be contributing to theoretical developments themselves. Since the Canadian entomologists also happened to believe in an ecological theory quite opposed to that supported rather more pragmatically by the entomologists in Riverside and Berkeley, they could not but have had grave misgivings about the work of the latter group.

These two groups of entomologists were mired in the difficulties involved in answering one of the most contested questions in ecology: how is the abundance of animal populations determined? As we have seen, there was no single answer to this question. They were both contributing quite significantly, albeit in very different ways, to two competing theories about the ecology of animal populations enjoying equal standing among ecologists at large. The empirical grounds for choosing between these two theories were, to say the least, far from conclusive; choices were strongly determined by prevailing institutional circumstances, and debate was couched in a mixed rhetoric linking views of nature and institutional considerations quite inextricably.

FROM INTEGRATED CONTROL TO INTEGRATED PEST MANAGEMENT

Notwithstanding their disagreements over the exact relationship between ecological theory and the work of biological and integrated control, by the mid-1960s, the entomologists in Riverside and Berkeley, as well as their counterparts in Canada, had developed the idea of integrated control into a major new way of thinking about the problems of crop protection. They all realized that these problems had to tackled from a far broader perspective than usual, one which would take into account the ecology of the entire crop and the impact of human activity on this ecology. The entire ecosystem had to be managed, and integrated control had to become integrated pest management. Transforming this conceptual advance into a practically significant one, however, was a far more difficult enterprise.

Most of the entomologists in Canada showed little interest in matching the conceptual advance with an institutional one. They had never been expected to tackle operational problems; their interest in questions of agricultural entomology was limited to the underlying scientific problems. Furthermore, as changes in the organization of agricultural research in Canada resulted in both the gradual re-orientation of work in federal laboratories toward more mission-oriented work and the transfer of responsibilities for research to the universities, most of these entomologists moved to university positions. There, they began to focus on altogether different problems. Donald Chant, for example, became closely involved with general environmental problems and the establishment of the Pollution Probe Foundation, while Ken Watt and Buzz Holling became interested in problems of environmental and urban planning. The task of transforming integrated control into a practical alternative to the current, economically and ecologically disastrous use of chemical insecticides was left therefore to the entomologists in Riverside and Berkeley. This parting of ways effectively ended the debate between the two groups, an ending which suggests that the notion of closure in scientific debate as a coming together of minds might need some reconsideration.[14]

I examined the difficulties involved in this transformation of integrated control by studying the competition between, on the one hand, the entomologists in Riverside and Berkeley, and, on the other, the proponents of a second alternative, which was developed by the Agricultural Research Service of the United States Department of Agriculture, under the leadership of Edward Knipling. The late 1950s and '60s marked a period of great division between American entomologists committed to the further development and improvement of chemical insecticides, and that minority interested in more ecologically sophisticated approaches. John Perkins has discussed this rift by focusing on its ideological roots. From the perhaps advantageous, but more likely, problematic, position of an actor in the very events he recounts, he argues that one group of entomologists believed that humans stood outside nature, and from there should strive for maximum control over nature. The release of sterile males to eradicate insect populations proposed by Knipling supposedly fell in this camp. The opposing group believed that humans were an integral part of nature, and should strive for a balance between their needs and those of other species. For Perkins, it was the latter view of nature that informed the development of integrated pest management led by the entomologists at Riverside and Berkeley. As we have seen, they believed quite strongly in the ability of predators and parasites to reduce populations of noxious insect species to such low levels that it was possible to imagine that the use of chemical insecticides might eventually be eliminated altogether. Since, in Perkins' view, these ideologies were fundamentally irreconcilable, they led inevitably to reciprocal and uncompromising rejection of the strategies proposed by each of the two groups.

It is interesting, however, that Rachel Carson, like Perkins, divided the world of crop protection along very similar ideological lines. In her view, one group of entomologists strove for absolute control over nature regardless of the ecological implications of their actions; she believed that this was an ideology born in the "Neanderthal age of philosophy and biology". To this group, she counterposed another one, whose members sought to replace classic chemical control with other strategies that were more attuned to the complex and dynamic nature of pest populations. Carson suggested that,

> through all these new, imaginative, and creative approaches to the problem of sharing our earth with other creatures there runs a constant theme, the awareness that we are dealing with life — with living populations and all their pressures and counterpressures, their surges and recessions. Only by taking account of such life forces and by cautiously seeking to guide them into channels favorable to ourselves can we hope to achieve a reasonable accommodation between the insect hordes and ourselves.[15]

In Perkins' view, this ecological perspective on pest control practice was peculiar to integrated pest management alone. It is ironic, then, and telling against the effectiveness of Perkins' explanatory model, that Carson included Knipling's plans to use sterile males to eradicate insect populations among these "new, imaginative and creative approaches".[16]

Just as Michael Smith has done for the now classic confrontation of environmentalist mythology, the clash between Gifford Pinchot and John Muir over the water reservoir at Hetch Hetchy, the successful development of integrated pest management cannot be viewed solely as a triumph of environmentalist ideology.[17] Rather, it must be viewed as the outcome of institutional and political conflicts, in which the entomologists were actively involved, sometimes using the rhetoric of environmentalism as one among many useful resources to advance their own goals. More specifically, the conflict between the entomologists in the United States Department of Agriculture and those in Berkeley and Riverside arose within the context of that antagonism between federal and local authority which has characterized so much of the history of the United States, even the history of American science.[18] Relationships between the Department and state agricultural experiment stations such as those in Riverside and Berkeley, almost directly reflecting this historical antagonism, have not always been easy, and they were particularly strained the 1960s when the proponents of integrated control and the sterile male release approach began to compete for limited public support. At about the same time, there was a significant shift in the American political and ideological landscape, a shift toward a lessened role for the federal government, to the supposed benefit of the states. The Nixon Administration played a major role in this political and ideological development, by harnessing to this end the political momentum of the environmentalist movement. The proponents of integrated pest management exploited this favorable political environment very adroitly.[19] They took advantage of their favored institutional location in the state agricultural experiment stations, and aligned themselves with environmentalist movement, adopting much of its rhetoric, to find a place in one of the most significant endeavors of the 1960s and '70s to resolve the problems of environmental degradation, the International Biological Program. Accomplishing this feat also meant, however, emphasizing an aspect of their work which had always taken the back-seat to their practical responsibilities. They had to become, at least superficially, "applied ecologists" in the strict sense used by their erstwhile Canadian critics.

INTEGRATED PEST MANAGEMENT AND ECOLOGY

The Canadian entomologist Donald Chant has suggested that many entomologists in the United States have lived their professional lives juggling and alternately donning two hats. They have worn either the hat of the insect ecologist, or that of the entomologist interested in finding the best way to eliminate a noxious insect species. The development of integrated control illustrated this point quite well. On the one hand, this novel strategy was supposed predicated on a keen knowledge of insect life histories and population dynamics; this was the insect ecologist speaking. On the other hand, its proponents also argued that its future development was predicated on the emergence of new, narrower spectrum insecticides; this was the view of entomologists interested in immediate solutions to the problems of their agricultural clientele. To a great extent, where the balance in this hat juggling exercise lay depended on the availability of the usually limited resources to pursue the first option. Note however that the tendency to pursue more explicitly ecological research was tempered by the prevailing ideology of service to the agricultural community, which sometimes subtly infiltrated entomologists' epistemology.

By the late 1960s, Carl Huffaker, from the division of Biological Control in Berkeley, had succeeded in integrating a large-scale national project on integrated pest management in the International Biological Program. The Program was, perhaps, the climactic expression of a novel approach to the solution of ecological problems, which had emerged during the 1950s and early '60s: systems ecology. The national integrated pest management project was viewed by many systems ecologists as a test of their approach to ecology to resolve pressing environmental problems, such as the excessive and mis-directed use of chemical insecticides. Systems ecology was, however, predicated on assumptions about the organization of biological communities that were in many ways antithetical to those informing the work of the entomologists in Riverside and Berkeley. On the other hand, the exact relationship between systems theory, the strongly positivistic theoretical framework informing the organization of ecological data characterizing systems ecology, and the collection of this data was sufficiently unclear that systems ecology could accommodate the conflicting ecological views of Huffaker and his associates as well as those of their erstwhile Canadian critics, new colleagues, and articulate exponents of systems ecology, Ken Watt and Buzz Holling. But, for those students at Berkeley, such as Andy Gutierrez, who learned about the relationship between ecology and integrated pest management during the evolution of the national integrated pest management project, the association was neither, as Paul Forman would put it, an institutional or a conceptual accommodation.[20] It probably was such an accommodation for Gutierrez' teacher, Huffaker, but for

Gutierrez himself, integrated pest management had to be advanced by the direct application of the best science known to him, systems ecology.

Not all ecologists, however, enjoyed the benefits from the rapid institutional expansion of the systems ecology during the 1960s. Theoretical population ecologists had been unwilling to associate the future of ecology with problems of environmental management because ecology had only recently begun to shed what was to them a disreputable image as the science of fisheries, wildlife and grassland management. They moved therefore to challenge the intellectual foundations of systems ecology. Richard Levins, at Harvard University, in particular, focused his criticism on those systems ecologists who had sought to bring systems theory to bear on ecological problems which theoretical population ecologists felt were their own exclusive domain, namely Watt, Holling, and all those who followed them. He argued that the strong empiricism characterizing their approach could not be anything but the result of excessive concern to solve immediate practical problems. This view seemed to be confirmed by the institutional affiliations of a large number of the ecologists involved with the International Biological Program: they held positions in departments of wildlife, fisheries, and grassland management in land-grant universities.[21] Furthermore, in the earlier years of their careers, Watt and Holling were associated with the Canadian Department of Agriculture, and were then avowedly interested in problems of environmental and urban management. All this confirmed an association of systems ecology and concern for practical issues, which according to Levins was extremely counterproductive.

In the late 1970s, Gutierrez responded to this criticism of the place of systems analysis in ecology, arguing that the criteria used by Levins (and by other theorectical population ecologists like him) to judge the value of different approaches to answering questions about ecological organization were more appropriate to physics than ecology. In his defense of a more empiricist approach to answer such questions, Gutierrez aligned himself with other critics of the contemporary development of ecology, such as George Oster and Donald Strong. For a brief moment he even seemed to align himself with Strong's colleague and most radical critic of this development, Daniel Simberloff. Simberloff claimed that ecologists should not start their work by presuming the existence of general patterns and then designing experiment to prove their existence, but instead should first prove that nature is not intrinsically idiosyncratic.[22] It may just be possible, Simberloff argued, that there can be no ecological counterparts to physicists' universal laws. Similarly, in the mid-1970s, Gutierrez suggested that the persistent difficulties involved in formulating general ecological statements that were supported by field data could be taken to mean that nature is fundamentally idiosyncratic. Within the next ten years, however, he withdrew from this position. The world was still for him far more complicated than some theoretical population ecologists would have liked to believe, but he now believed that some day ecologists would be able to formulate general statements about the organization of nature, about how the "balance of nature" is maintained. Gutierrez could not have pursued his earlier views any further than he had. It would have condemned him to an exile on the periphery of ecology, leaving him with no institutional home within entomology, because his was no longer a discourse about how to better protect agricultural crops against insect pests.

Like the entomologists in the Research Branch of the Canadian Department of Agriculture during the 1960s, Gutierrez, and other entomologists who had learned their trade during the development of integrated pest management projects, such as Pedro Barbosa, at the University of Maryland, and Robert Luck, at the University of California in Riverside, had finally moved economic entomology to the frontiers of contemporary ecology, as Rachel Carson had recommended some thirty years earlier. Their work was "applied ecology" in the strictest sense, almost matching the mythological relationship between particle physics and nuclear engineering.[23] However influenced intellectually by their past associations with work of crop protection, Gutierrez, Luck, and Barbosa now spoke to a world far removed from that of farmers and environmentalists interested in more effective and less environmentally disruptive means of controlling insect pests.[24] Ecologists had become their prime audience, and it was the reward structures of ecology rather than economic entomology which would determine their professional future. In other words, this transformation closely parallels that which John Servos has observed in his more limited study of federal funding for engineering and physical sciences, and its effects on the relationship between these sciences at the Mellon Institute.[25] Turning things on their head, this study of the life sciences and agriculture could provide a framework to help us make sense of the relationship between particle physics and nuclear engineering, or science and technology in North America more generally. As I suggested in the introduction, relationships between ideas, such as science and technology, are relationships between the people who hold them. Alliances between them, and hence, between their ideas, are determined by their negotiations within, and between, the networks of institutions in which they participate, ranging from ever evolving and changing scientific disciplines, such as entomology and ecology, to the governmental and industrial organizations that sometimes support these disciplines' growth for quite practical reasons. The relationships between these ideas and institutions are not, however, straightforward mappings of the latter onto the former. This is because the people involved are constantly and unevenly mobilizing one or the other, or a mixture of both, in other words, they are constantly engaging in what John Law has called "heterogeneous engineering", to advance their own careers.[26] In the process, their institutions and ideas are constantly and unevenly changed: in building alliances with ecologists, the entomologist considered here changed both themselves and the intellectual foundations of ecology.

FINAL REMARKS

The publication of *Silent Spring* and the subsequent growth of public concern about issues of environmental degradation were crucially important for the establishment of institutional support to develop integrated pest management into a viable alternative to the contemporary misuse of chemical insecticides. Promoting this more ecologically sophisticated approach to crop protection was but one response to public demands for greater social and environmental responsibility in the domain of agricultural policy.[27] Integrated pest management was, and still is considered an important element in a new approach to agricultural production that is more humane and more attuned to the view that agriculture should be something more than just another aspect of a capitalist economy.[28] The fact that integrated pest management appears to have failed to achieve these utopian goals cannot just be attributed, as some entomologists in the Division of Biological Control at Berkeley have suggested in recent years, to corruption of the original intentions of its proponents, or to the corrupting action of industrial interests. In 1964 Paul De Bach suggested that,

> the most practical solution [to the problems of pest control] will be a thoroughly scientific integration of chemical and biological control. The trend to larger and larger farming operations will make integrated and supervised pest control more feasible for several reasons. Large-scale farming producers tend towards uniformity of habitats and ecological interactions which make the development and application of manipulation and integration programs easier and more reliable than if many small farmers were involved. Further, the extension of ideas and practices is much easier with one large operator than with 20 small ones, and large-scale operators often more easily recognize the need for sound advice and consequently are more likely to support a well-trained supervising entomologist.[29]

Suffice it to say that the "supervising entomologist" was to be an insect ecologist. Even in this very early formulation, integrated pest management was envisioned as an element in the establishment of agricultural productivity in a more capitalistically rational framework, and insect ecologists were to play an active role in this transformation. Unless paid for by the taxpayer, the advice of farm management consultants is, for farmers and growers, an even more expensive capital input than chemical insecticides; integrated pest management in California was, is, part of "big farming". The high cost of this technology is also motivating current criticism of the efforts by the World Bank and the Consultative Group for International Agricultural Research to promote the use of integrated pest management in agricultural development schemes for the lesser developed countries as quite inappropriate for the most agricultural producers in these countries.[30]

More significantly, insect ecologists' bid for a place in the evolution of big farming faltered during the 1980s, as a different group of entomologist reaped the benefits of the further advancement of capitalism in agriculture. The neo-federalist program of the Nixon Administration, which had helped the development of integrated pest management so much during the 1970s, also aimed to reduce federal funding for all areas of agricultural research except those that were so removed from the development of commercially valuable products that the private sector would be unwilling to support them. Not all areas of research could promise to eventually produce something equally attractive to private investors. More research on the ecological foundations of integrated pest management might facilitate the work of private farm management consultants, but would not produce much more. The controversial *Report of the Committee on Research Advisory to the United States Department of Agriculture*, the "Pound Report" suggested instead that the Department of Agriculture should encourage more work in molecular biology. This was, after all, the truly leading edge of biology, and, moreover, the practical applications of this science promised to be quite lucrative to chemical and pharmaceutical industry. Enticed by the prospect of developing plants which could produce viral compounds toxic to insect species attacking them, or, more significantly, by the prospect of developing plants resistant to particular herbicides, these industries began to establish closer links with researchers in federal and state agricultural research organizations.[31] Repeating the pattern of earlier years, this support for research in molecular biology encouraged a shift away from work on the ecological foundations of integrated pest management. Integrated pest management came to mean the optimal management of more sophisticated chemical insecticides and the introduction of crops genetically engineered for greater resistance to both noxious insect species and these more sophisticated insecticides.[32] This shift motivated entomologists from the Division of Biological Control at Berkeley to challenge the definition of biological control proposed in 1987 by the National Academy of Sciences in its *Report of the Research Briefing Panel on Biological Control in Managed Ecosystems.* This definition emphasized the role of molecular biology and genetic engineering in the future of biological control and integrated pest management to such an extent that they felt that it debased the importance of research in "population dynamics, evolutionary ecology, and systematics".[33] The seriousness of this threat was epitomized one year later by the closure of all remnants of the Division of Biological Control in Riverside. In Berkeley, the Division still maintains, rather tenuously, a separate identity from Department of Entomology, but now work there is not very dissimilar from that of ecologists interested in, for example, the organization of tropical ecosystems. What are the costs of this change? Like that of the Canadian entomologists before them, the work of these American entomologists is having only a very minimal impact on what happens in fields and orchards. American farmers ceased to be the audience of entomologists like them a long time ago. They heed, instead, the advice of salespeople from Monsanto, who provide them with the latest products of molecular biology and genetic engineering. Environmentalists have turned their attention to the concept of a "sustainable agriculture", which will be far less intensive than that envisioned by De Bach in 1964. Thus, David Pimentel has recently written that,

> progress has been made on pesticide problems, but *Silent Spring* is not entirely behind us. Pesticide use continues, and the quantities of pesticides applied grow annually despite support for [integrated pest management] control. In future decades, as the

> world population grows rapidly and agricultural production is stretched to meet food needs, we should not forget Carson's warnings.
>
> Pesticides will continue to be effective pest controls, but the challenge now is find ways to use them judiciously to avoid the environmental hazards and human poisoning that exist today. With this goal for research and development we can achieve, relatively safe pest control programs.[34]

Contrary to Pimentel's assessment, however, the answers to the growing use of pesticides and the environmental problems caused by it will not emerge solely from the future conceptual advances in entomology and ecology he and many, many other entomologists and ecologists had sought over the previous century. All scientific discourse is embedded in a social one, and any discussion of the social problems arising from technological change cannot ignore the active and powerful role of scientists in shaping these problems. There are lessons here for entomologists, environmentalists and, not least, for historians.

NOTES

1 President's Science Advisory Commitee, *Restoring the Quality of our Environment*, (Washington: United States Government Printing Office, 1965), p. 41.

2 R.P. McIntosh, *The Background to Ecology. Concept and Theory* (Cambridge: Cambridge University Press, 1985), p. ix.

3 See G.N. Gilbert & M. Mulkay, *Opening Pandora's Box: A Sociological Analysis of Scientists' Discourse* (Cambridge: Cambridge University Press, 1984).

4 C. Sorensen, *Brethren of the Net: American entomology, 1840–1880* (Ph.D. dissertation, University of California, Davis, 1984), pp. 164–89; and Perkins, *Insects, Experts and the Insecticide Crisis*, pp. 209–64.

5 M. Nicholson, "National styles, divergent classifications: a case study from the history of American and French plant ecology" in L. Hargens *et al.* (eds), *Knowlegde and Society: Studies in the Sociology of Science, Past and Present* (Greenwich: JAI Press, 1989), pp. 139–86; and J. Harwood, "National styles in science: genetics in Germany and the United States between the world wars", *Isis* 78 (1987): 390–414.

6 See T.R. Mitchell & R. Lowe, "To sow contentment: philanthropy, scientific agriculture, and making of the New South, 1906–1920", *Journal of Social History* 24 (1990): 317–340; and D. Danbom, *The Resisted Revolution: Urban American and the Industrialization of Agriculture, 1900–1930* (Ames: Iowa State University Press, 1979).

7 W.M. Wheeler quoted M.A. Evans & H.E. Evans, *William Morton Wheeler, Biologist* (Cambridge: Harvard University Press, 1970), p. 180.

8 D. Fitzgerald, *The Business of Breeding: Hybrid Corn in Illinois, 1890–1940* (Ithaca: Cornell University Press, 1990); and R.C. Tobey,

Saving the Prairies: The Life Cycle of the Founding School of American Plant Ecology, 1895–1955 (Berkeley: University of California Press, 1981).

9 M. Dubasak, *Wilderness Preservation: A Cross-Cultural Comparison of Canada and the United States* (New York: Garland Press, 1990).

10 On the organization of British agriculture research, and its implications for the conduct of research, see P. Palladino, "Science and the economy: plant breeding research in Great Britain, 1920–70", *Economic History Review* (forthcoming).

11 Quoted in W. Downes, "Fifty years of entomology on Vancouver Island", *Proceedings of the Entomological Society of British Columbia* 48 (1951): 9–16, on p. 11.

12 Paul De Bach (now deceased) to Paolo Palladino, 9 July 1987.

13 C.B. Huffaker *et al.*, "The natural enemy component in natural control and the theory of biological control" in Huffaker (ed.), *Biological Control* (New York: Plenum Press, 1971), pp. 16–67 on p. 46.

14 See M. Rudwick, "The closure of the Devonian controversy" in *British Society for the History of Science and the History of Science Society: Program, Paper, and Abstracts for Joint Conference* (Madison: Omnipress, 1988), pp. 155–9.

15 R. Carson, *Silent Spring* (Boston: Houghton Mifflin Company, 1987), p. 296.

16 Carson, *Silent Spring*, pp. 277–297.

17 M.L. Smith, *Pacific Visions: California Scientists and the Environment, 1850–1915* (New Haven: Yale University Press, 1987), pp. 166–85.

18 See Fitzgerald, *Business of Breeding*, pp. 63–9; and R.C. Sawyer, "Monopolizing the insect trade: biological control in the USDA, 1881–1951", *Agricultural History* 64 (1990): 271–85.

19 For another example of how scientists during this period exploited pressing social issues, such as growing public fear of cancer, for their own interest in institutional expansion, see M. Heirich, "Why we avoid the key questions: how shifts in the funding of scientific inquiry affect decision making about science" in D.A. Jackson & S.P. Stich (eds), *The Recombinant DNA Debate* (Eaglewood Cliffs: Prentice-Hall, 1979), pp. 234–60.

20 P. Forman, "Weimar culture, causality, and quantum physics: Adaptation by German physicists and mathematicians to a hostile intellectual environment", *Historical Studies in the Physical Sciences* 3 (1971): 1–116.

21 See C. Kwa, *Mimicking Nature: the Development of Systems Ecology in the United States, 1950–1975* (Ph.D. dissertation, University of Amsterdam, 1989), pp. 100 and 117–8; and J.B. Hagen, *An Entangled Bank: The Origins of Systems Ecology* (New Brunswick: Rutgers University Press, 1993), pp. 189–97.

22 D. Simberloff, "A succession of paradigms in ecology: essentialism to materialism and probabilism" in E. Saarinen (ed.), *Conceptual Issues in Ecology* (Dordrecht: Reidel, 1982), pp. 63–99.

23 On the idea of "technology" as "applied science", and its problems, see J.M. Staudenmaier, *Technology's Storytellers: Reweaving the Human Fabric* (Cambridge: MIT Press, 1985), pp. 83–120.

24 See P. Barbosa & V.A. Krishchnik, "Influence of alkaloids on feeding preference on eastern deciduous trees by the Gypsy Moth", *American Naturalist* 130 (1987): 53–69; and R.F Luck & H. Podler, "Competitive exclusion of *Aphytis lignanensis* by *A. melinus*: potential role of host size", *Ecology* 66 (1985): 904–13.

25 J.W. Servos, "Changing partners: the Mellon Institute, private industry, and the federal patron", *Technology and Culture* 35 (1994): 221–57.

26 J. Law, "Technology and heterogeneous engineering: the case of Protuguese expansion" in W.E. Bijker *et al.* (eds), *The Social Construct of Technological Systems* (Cambridge: MIT Press, 1987), pp. 111–34.

27 See A. Mayer & J. Mayer, "Agriculture, the island empire", *Daedalus* 103 (3) (1974): 83–95.

28 See P. Weber, "A place for pesticides", *World Watch* 5 (1992): 18–25; and also R. Levins, "Perspective in integrated pest management: from an industrial to an ecological model of pest management" in Kogan, *Ecological Theory and Integrated Pest Management Practice*, pp. 1–18.

29 P. De Bach, "Successes, trends, and future possibilities", in De Bach (ed.), *Biological Control of Insect Pests and Weeds* (New York: Reinhold Publishing, 1964), pp. 673–713, on p. 713.

30 S. Morse, Discussion Paper, Institute of Development Studies, University of East Anglia, January 1995.

31 F.H. Buttel, "Biotechnology and agricultural research policy: emergent issues" in K.A. Dahlberg (ed.), *New Directions for Agriculture and Agricultural Research* (Totowa: Rowman & Allanheld, 1986), pp. 312–47.

32 See R. Garcia & D.L. Dahlsten, "Pest control strategies" in G. DeBell (ed.), *The New Environmental Handbook* (San Francisco: Friends of the Earth, 1980), pp. 196–209; see also E. Auchard, "Back to basics: progressive researchers learn from old-style farming", *Berkeley Graduate* 6 (1992): 14–21.

33 See R. Garcia *et al.*, "Comments on a redefinition of biological control", *BioScience* 38 (1988): 692–4.

34 Pimentel, "Is *Silent Spring* behind us?" in G.J. DeMarco *et al.* (eds), *Silent Spring* revisited (Washington: American Chemical Society, 1987), pp. 173–187, on pp. 184–5.

Index